SpringerBriefs in Applied Sciences and Technology

SpringerBriefs present concise summaries of cutting-edge research and practical applications across a wide spectrum of fields. Featuring compact volumes of 50 to 125 pages, the series covers a range of content from professional to academic.

Typical publications can be:

- A timely report of state-of-the art methods
- An introduction to or a manual for the application of mathematical or computer techniques
- A bridge between new research results, as published in journal articles
- A snapshot of a hot or emerging topic
- An in-depth case study
- A presentation of core concepts that students must understand in order to make independent contributions

SpringerBriefs are characterized by fast, global electronic dissemination, standard publishing contracts, standardized manuscript preparation and formatting guidelines, and expedited production schedules.

On the one hand, **SpringerBriefs in Applied Sciences and Technology** are devoted to the publication of fundamentals and applications within the different classical engineering disciplines as well as in interdisciplinary fields that recently emerged between these areas. On the other hand, as the boundary separating fundamental research and applied technology is more and more dissolving, this series is particularly open to trans-disciplinary topics between fundamental science and engineering.

Indexed by EI-Compendex, SCOPUS and Springerlink.

Manish Kumar Goyal · Shivukumar Rakkasagi

Combating Forest Loss and Fires

Technology–Policy Approaches in Key Biodiversity Areas

Manish Kumar Goyal ⓘ
Department of Civil Engineering
Indian Institute of Technology Indore
Indore, Madhya Pradesh, India

Shivukumar Rakkasagi ⓘ
Department of Civil Engineering
Indian Institute of Technology Indore
Indore, Madhya Pradesh, India

ISSN 2191-530X   ISSN 2191-5318   (electronic)
SpringerBriefs in Applied Sciences and Technology
ISBN 978-3-031-84819-3   ISBN 978-3-031-84820-9   (eBook)
https://doi.org/10.1007/978-3-031-84820-9

© The Author(s), under exclusive license to Springer Nature Switzerland AG 2025

This work is subject to copyright. All rights are solely and exclusively licensed by the Publisher, whether the whole or part of the material is concerned, specifically the rights of translation, reprinting, reuse of illustrations, recitation, broadcasting, reproduction on microfilms or in any other physical way, and transmission or information storage and retrieval, electronic adaptation, computer software, or by similar or dissimilar methodology now known or hereafter developed.

The use of general descriptive names, registered names, trademarks, service marks, etc. in this publication does not imply, even in the absence of a specific statement, that such names are exempt from the relevant protective laws and regulations and therefore free for general use.

The publisher, the authors and the editors are safe to assume that the advice and information in this book are believed to be true and accurate at the date of publication. Neither the publisher nor the authors or the editors give a warranty, expressed or implied, with respect to the material contained herein or for any errors or omissions that may have been made. The publisher remains neutral with regard to jurisdictional claims in published maps and institutional affiliations.

This Springer imprint is published by the registered company Springer Nature Switzerland AG
The registered company address is: Gewerbestrasse 11, 6330 Cham, Switzerland

If disposing of this product, please recycle the paper.

Preface

Key Biodiversity Areas (KBAs) are sites that contribute significantly to the global persistence of biodiversity. These sites with geographically defined boundaries hold biodiversity (ecosystems, species, or genes) of international significance. Forest loss within KBAs is a pressing concern, as deforestation leads to habitat destruction and fragmentation. The book delves into the causes and consequences of forest cover loss in KBAs, using open-source datasets and methodologies to quantify the extent of the problem. By highlighting the technology–policy interface, it suggests ways to combat deforestation effectively. Forest fires pose another significant threat to KBAs. The book examines the frequency and intensity of fires in these areas, discussing their impact on biodiversity and ecosystem services. It outlines strategies for fire prevention and management, emphasizing the role of technology in understanding and mitigating the impacts on forest biodiversity, an often-overlooked issue. By analyzing stable night lights in KBAs, the book sheds light on how human development encroaches upon natural habitats and proposes measures to mitigate its impact.

The book emphasizes the importance of integrating technology with effective policy-making. It demonstrates how remote sensing and data analytics can inform conservation strategies and policy decisions. This approach offers a blueprint for evidence-based conservation practices that can be applied beyond India. The book also outlines future directions for KBA monitoring, highlighting emerging technologies and their potential applications in conservation. By doing so, the book identifies problems and provides a roadmap for solutions. This work aims to bridge the gap between scientific knowledge and policy implementation, offering practical strategies for protecting KBAs while supporting sustainable development.

The target audience for this book includes scientists, conservation biologists, environmental policymakers, sustainable development practitioners, and ecology and environmental science researchers. It will also be valuable for NGOs working in biodiversity conservation, government agencies responsible for environmental management, and students pursuing advanced studies in related fields.

Indore, India

Manish Kumar Goyal
Shivukumar Rakkasagi

Contents

1	**Key Biodiversity Areas (KBAs)**		1
	1.1	Background	1
	1.2	Definition and Relevance of KBAs to SDGs	2
		1.2.1 Definitions of KBAs	2
		1.2.2 Relevance of KBAs to Sustainable Development Goals (SDGs)	2
	1.3	Spatial Distribution of KBAs: Global and India	5
		1.3.1 KBAs Distribution Around the Globe	5
		1.3.2 KBAs Distribution Across India	7
	1.4	Criteria and Thresholds for Selecting KBAs	9
		1.4.1 Criteria for Selecting KBAs	9
		1.4.2 Thresholds for Selecting KBAs	10
		1.4.3 Application and Considerations of Criteria and Thresholds	10
	1.5	Rationale for Selecting KBAs	11
	1.6	Importance of Monitoring KBAs	12
	1.7	Conclusion	13
	References		14
2	**Key Biodiversity Areas and Forest Cover Loss**		17
	2.1	Introduction	17
	2.2	Impacts of Deforestation on Biodiversity	18
	2.3	Datasets and Methodology for Forest Cover Losses	20
		2.3.1 Data Sources and Preprocessing	20
		2.3.2 Google Earth Engine Implementation	20
		2.3.3 Statistical Analysis	21
		2.3.4 Spatial and Temporal Analysis	21
		2.3.5 Validation and Uncertainty Assessment	21
		2.3.6 Limitations and Considerations	22
	2.4	Results and Discussion of Forest Loss in Indian KBAs	22
		2.4.1 Overall Trend Analysis	22

		2.4.2 Regional Patterns	23
		2.4.3 Implications and Discussion	25
		2.4.4 Limitations and Future Research Directions	27
	2.5	Technology–Policy Interface to Reduce Deforestation	28
	2.6	Conclusion	30
	References		31
3	**Key Biodiversity Areas and Forest Fire**		**35**
	3.1	Introduction	35
	3.2	Impacts of Forest Fires on KBAs	36
	3.3	Datasets and Methodology for Forest Fire Analysis	38
		3.3.1 Data Sources and Preprocessing	38
		3.3.2 Google Earth Engine Implementation	38
		3.3.3 Statistical Analysis	39
		3.3.4 Spatial and Temporal Analysis	39
		3.3.5 Validation and Uncertainty Assessment	40
		3.3.6 Limitations and Considerations	40
	3.4	Results and Discussion of Forest Fire in Indian KBAs	41
		3.4.1 Overall Trend Analysis	41
		3.4.2 Regional Patterns	41
		3.4.3 Implications and Discussion	44
		3.4.4 Limitations and Future Research Directions	45
	3.5	Technology–Policy Interface to Prevent and Manage Forest Fires	46
	3.6	Conclusion	49
	References		49
4	**Key Biodiversity Areas and Stable Night Lights**		**53**
	4.1	Introduction	53
	4.2	Effects of Artificial Light on Wildlife	55
	4.3	Datasets and Methodology for Monitoring Night-Time Lighting	56
		4.3.1 Data Sources and Preprocessing	56
		4.3.2 Google Earth Engine Implementation	57
		4.3.3 Statistical Analysis	57
		4.3.4 Spatial and Temporal Analysis	57
		4.3.5 Validation of the Results	58
		4.3.6 Limitations and Considerations	58
	4.4	Results and Discussion of Night-Time Lighting in Indian KBAs	59
		4.4.1 Overall Trend Analysis	59
		4.4.2 Regional Patterns	60
		4.4.3 Implications and Discussion	62
		4.4.4 Limitations and Future Research Directions	63
	4.5	Technology–Policy Interface to Limit the Spread of Stable Night Lights	64
	4.6	Conclusion	66
	References		67

5	**Policy Recommendations and Future Directions**		71
	5.1 Achievement of the SDGs		71
		5.1.1 SDGs for Forest Cover Change	71
		5.1.2 SDGs for Forest Fire Trends	73
		5.1.3 SDGs for Night-Time Lighting Trends	74
	5.2 Future Technological Advances in KBA Monitoring		75
	5.3 Policy Recommendations for KBA Protection		78
	5.4 Conclusion		82
	References		83

Chapter 1
Key Biodiversity Areas (KBAs)

1.1 Background

The *International Union for Conservation of Nature (IUCN) Red List of Threatened Species* develops quantitative criteria to evaluate extinction threats at the species level [1]. Biodiversity is endemic to a particular geographical area, biological processes, ecological integrity, endangered biodiversity, and irreplaceability [2]. The *IUCN* advocates for Key Biodiversity Areas (KBAs) to recognize sites that hold significant importance for the long-term survival of biodiversity globally [1]. The global standard for KBA recognition was agreed in 2016, and more than 16,000 KBAs have already been recognized worldwide [3]. KBAs are important sites recognized for their substantial contributions to global biodiversity conservation. KBAs focus on protecting worldwide biodiversity and are acclaimed as vital land, forest, freshwater, and marine areas for threatened flora and fauna. KBAs are identified geographically and necessary for preserving diverse ecosystems and species of international importance [4]. KBAs are identified using globally standardized criteria established by the *IUCN* in 2016, and they extend the concept of Important Bird Areas (IBAs) to involve other taxonomic groups and ecosystems [1]. The *KBA Partnership* supports and promotes the identification and conservation of KBAs by bringing together most of the foremost global international conservation organizations [5] (please refer to link http://www.keybiodiversityareas.org/kba-partners).

1.2 Definition and Relevance of KBAs to SDGs

1.2.1 Definitions of KBAs

While there is a core consistency in the concept of KBAs, their specific phrase and prominence of their definitions can vary across different conservation organizations and reports. Here are some important definitions of how KBAs are described:

(a) **IUCN definition of KBAs (2007)** [6]: "KBAs are sites of global significance for biodiversity conservation. They are identified using globally standard criteria and thresholds, based on the needs of biodiversity requiring safeguards at the site scale. These criteria are based on the framework of vulnerability and irreplaceability widely used in systematic conservation planning."
(b) **IUCN definition of KBAs (2014)** [7]: "Key Biodiversity Areas (KBAs) are sites that contribute significantly to the global persistence of biodiversity. These are sites with geographically defined boundaries that hold biodiversity (ecosystems, species, or genes) of international significance. KBAs are endorsed by conservation organizations but do not necessarily have legal designation for any particular kind of land use (e.g., they are not necessarily protected areas)."
(c) **UN Environment Programme (UNEP) definition of KBAs (2014)** [8]: "Sites contributing significantly to the global persistence of biodiversity. They represent the most important sites for biodiversity conservation worldwide and are identified nationally using globally standardized criteria and thresholds."
(d) **Definition on the website of Nature Canada (2017)** [9]: "KBAs are sites that contribute to the global persistence of biodiversity. This means they are sites that are representative and significant, in one way or another, of the wide array of ecosystems, creatures, and species found around the world. The KBA designation covers areas important in terms of animal species but also extends to areas significant for their plant life or their life-sustaining environment."
(e) **Definition on the website of Marine Biodiversity Unit (2024)** [10]: "Key Biodiversity Areas (KBA) are sites contributing significantly to the global persistence of biodiversity, in terrestrial, freshwater and marine ecosystems. Sites qualify as global KBAs if they meet one or more of 11 criteria, clustered into five categories: threatened biodiversity; geographically restricted biodiversity; ecological integrity; biological processes; and, irreplaceability."

1.2.2 Relevance of KBAs to Sustainable Development Goals (SDGs)

KBAs play an essential role in achieving various SDGs due to their importance in biodiversity conservation and ecosystem services [11]. These areas directly support these goals by identifying and protecting critical habitats for diverse species in both aquatic and terrestrial ecosystems [12]. Many KBAs encompass essential carbon

1.2 Definition and Relevance of KBAs to SDGs

sinks and provide natural buffers against climate-related hazards [13], contributing to mitigation and adaptation strategies. These areas often include crucial freshwater ecosystems, contributing to water quality and quantity through natural filtration and flow regulation. KBAs support local livelihoods and food security by preserving ecosystems that provide essential services and maintaining genetic diversity necessary for agriculture [14]. Urban KBAs contribute to city sustainability and resilience by providing green spaces and supporting urban biodiversity [15]. KBAs highlight areas where sustainable resource management is crucial and provide opportunities for responsible consumption practices [16]. By identifying and protecting KBAs, we can progress toward multiple SDGs (Fig. 1.1), showcasing the interconnected nature of biodiversity conservation and sustainable development.

A brief explanation of KBAs relevance to specific SDGs is as follows:

(a) **SDG 2 (Zero Hunger)**: KBAs contribute to these goals by:

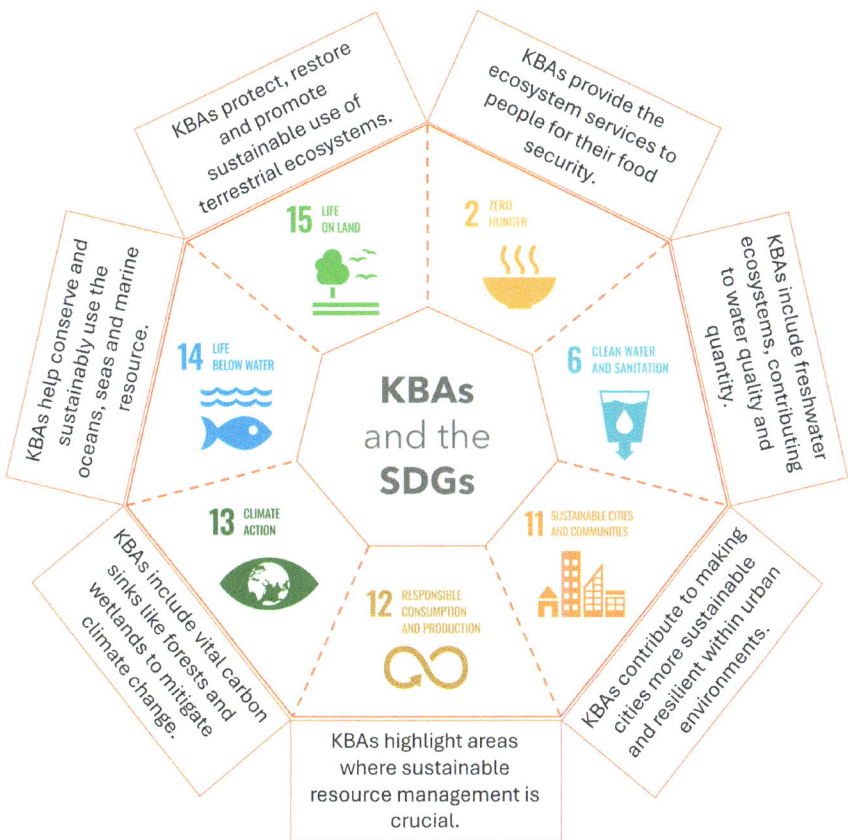

Fig. 1.1 Illustration of how the protection and management of KBAs directly contribute to multiple SDGs, emphasizing the crucial role of biodiversity conservation in sustainable development

- Preserving ecosystems that provide essential services (food, medicine, raw materials) to local communities.
- Maintaining the genetic diversity of wild relatives of crops is crucial for food security and agricultural resilience.
- Supporting sustainable livelihoods through eco-tourism and sustainable resource use.

Example: A coastal KBA can protect fish breeding grounds, supporting local fisheries and food security [17, 18].

(b) **SDG 6 (Clean Water and Sanitation)**: KBAs support this goal by:
- Encompassing watersheds and freshwater ecosystems are crucial for water supply.
- Maintaining natural water filtration systems improves water quality.
- Regulating water flow contributes to water quantity and availability.

Example: A KBA that includes a mountain forest might protect the watershed for a major river, ensuring a clean water supply for downstream communities [19].

(c) **SDG 11 (Sustainable Cities and Communities)**: Urban KBAs contribute by:
- Providing green spaces that improve air quality and offer recreational opportunities.
- Supporting urban biodiversity, enhancing city resilience and liveability.
- Offering nature-based solutions for urban challenges like flooding or heat island effects.

Example: An urban wetland KBA can provide flood protection, recreation area, and habitat for migratory birds [20].

(d) **SDG 12 (Responsible Consumption and Production)**: KBAs support this goal by:
- Highlighting areas where sustainable resource management is crucial.
- Providing opportunities for sustainable eco-tourism and non-timber forest products.
- Serving as living laboratories for developing sustainable production methods.

Example: A forest KBA can be managed for sustainable harvest of medicinal plants, promoting responsible use of natural resources [21].

(e) **SDG 13 (Climate Action)**: KBAs contribute to climate action by:
- Encompassing major carbon sinks like tropical rainforests, peatlands, and seagrass meadows.
- Protecting these carbon-rich ecosystems, thus mitigating climate change by reducing greenhouse gas emissions from land-use change.
- Preserving natural buffers against climate-related hazards (e.g., mangroves protecting coastlines from storm surges).

Example: Protecting a KBA in the rainforest preserves biodiversity and maintains a significant carbon sink [22].

(f) **SDG 14 (Life Below Water) and SDG 15 (Life on Land)**: KBAs are directly aligned with these goals as they:

- Identify critical habitats for biodiversity in both aquatic and terrestrial ecosystems.
- Provide a framework for prioritizing conservation efforts, helping to meet targets for protected area coverage.
- Support the preservation of endangered species by highlighting their critical habitats.
- Contribute to ecosystem restoration efforts by identifying degraded areas of high biodiversity value.

Example: Marine KBAs can include coral reefs or mangrove forests, which are crucial for numerous species and ecosystem services [23].

1.3 Spatial Distribution of KBAs: Global and India

1.3.1 KBAs Distribution Around the Globe

Understanding the distribution of KBAs across different geographical regions is vital for prioritizing conservation attempts, distributing resources effectively, and forming targeted strategies for biodiversity protection [22]. The global standard for KBA recognition was agreed in 2016, and more than 16,000 KBAs (exactly 16,428 as of July 30, 2024) have already been recognized worldwide [3]. The region-wise distribution of KBAs globally, as presented in Fig. 1.2, provides valuable insights into the spatial patterns and locations of biodiversity hotspots. Furthermore, we categorized KBAs into 13 major geographical regions to shed light on the global distribution of these critical global areas. The regions under consideration include Africa, Antarctica, Asia, Australasia, the Caribbean, Central America, Central Asia, Europe, High Seas, the Middle East, North America, Oceania, and South America (Fig. 1.3). In fact, Europe emerges as the region with the highest concentration of KBAs, count of 4862, leading with biodiversity richness, whereas Asia follows closely with 3427 KBAs, emphasizing its immense ecological importance as a biodiversity powerhouse. Interestingly, Asia has the largest total area of KBAs at about 3,461,777 km^2, substantially higher than any other region (Fig. 1.4). South America and Oceania follow as the second and third largest regions in terms of KBA area, with about 2,304,807 km^2 and 2,298,065 km^2, respectively. This distribution displays various factors, including the size of the regions, their biodiversity richness, and the extent of habitat conservation. The large areas in Asia, South America, and Africa align with major biodiversity areas on these continents.

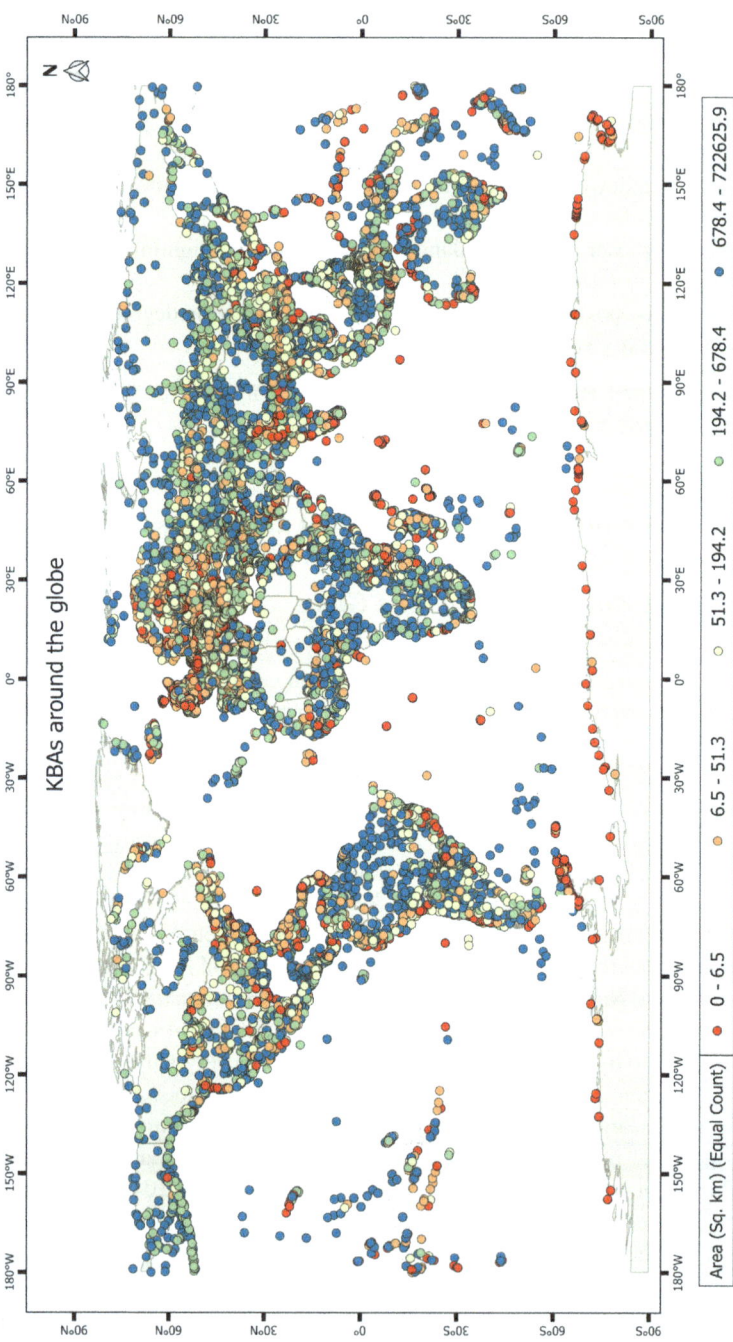

Fig. 1.2 Geographic distribution and location of KBAs across the globe. Most KBAs are located in Europe, parts of Africa, South and Southeast Asia, and parts of South and Central America, with the area ranging from less than 6.5 to over 722,000 km^2

1.3 Spatial Distribution of KBAs: Global and India

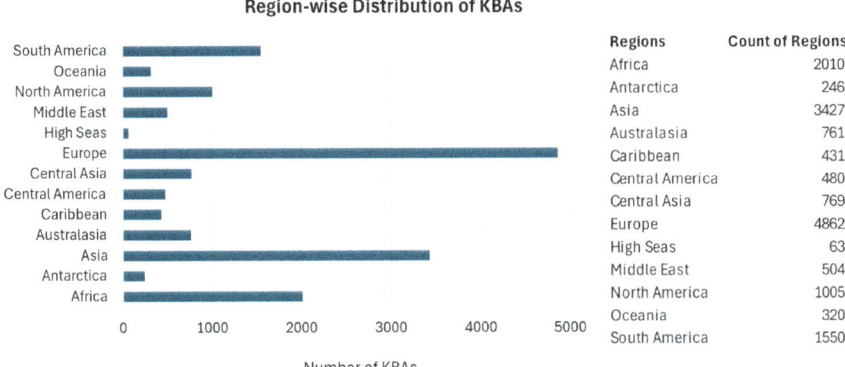

Fig. 1.3 Chart shows the global distribution of KBAs across 13 major geographical regions. The data is presented through a horizontal bar chart for visual comparison and a corresponding table with precise counts

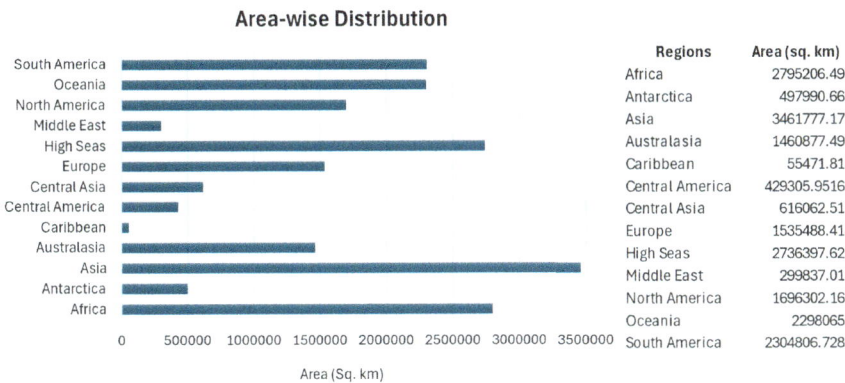

Fig. 1.4 Chart shows the surface area of KBAs across 13 major geographical regions. The data is presented through a horizontal bar chart for visual comparison and a corresponding table with a precise area in km^2

1.3.2 KBAs Distribution Across India

India is home to a high level of biodiversity due to its varied topography, diverse ecosystems, and climatic conditions [24, 25]. There are over 650 KBAs identified across 50 ecoregions in India, covering an area of over 10,600 km^2 (Fig. 1.5).

Some of the most notable KBAs in India include the Western Ghats, Eastern Himalayas, Sundarbans, Great Nicobar Biosphere Reserve, Gulf of Mannar, Kaziranga National Park, and Nanda Devi Biosphere Reserve. By focusing conservation efforts on these critical areas, India can maximize the benefits of biodiversity conservation while minimizing the negative impacts of development and other

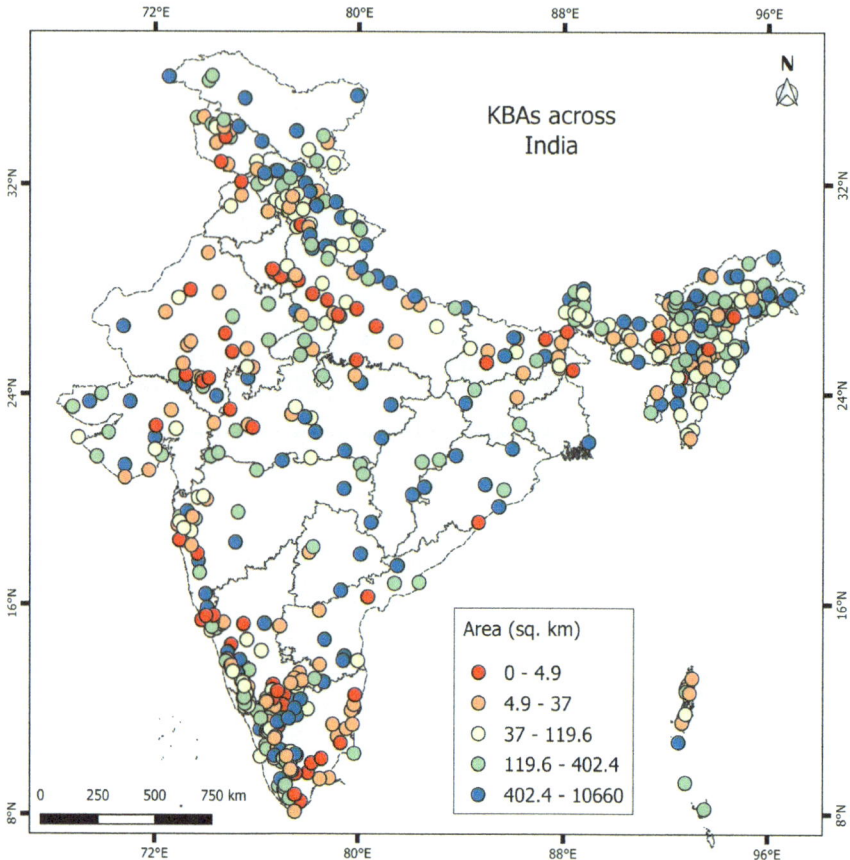

Fig. 1.5 Geographic distribution and location of KBAs across India. Most KBAs are located in the Northeast region, Western Ghats, and Andaman and Nicobar Islands, with the area ranging from less than 4.9 to over 10,600 km^2

human activities [26, 27]. KBAs in India are distributed across diverse landscapes, from the Himalayan peaks in the North to the lush tropical regions in the South and from the arid Western deserts to the verdant Eastern rainforests [28]. The Northeast, including states like Arunachal Pradesh, Assam, and Manipur, stands out as a significant biodiversity hotspot with numerous KBAs. Another biodiversity-rich region, the Western Ghats, is well-represented by KBAs throughout Kerala, Karnataka, and Tamil Nadu. The Andaman and Nicobar Islands also feature prominently, highlighting the ecological importance of these island ecosystems. India's largest KBA is the Jawaharlal Nehru Bustard Sanctuary, spanning an impressive 10,660.03 km^2. Other notable large-scale KBAs include the Desert National Park and Hemis National Park. Contrastingly, some KBAs are small, such as Telineelapuram, which covers a mere 0.12 km^2, showing the diverse nature of India's ecosystems and the varying sizes at which conservation efforts are focused.

1.4 Criteria and Thresholds for Selecting KBAs

KBAs are identified using standardized criteria to ensure that areas support significant species populations, maintain ecosystem integrity, and represent unique ecological regions. The KBA criteria and thresholds represent the advancement in standardizing the identification of globally significant sites for biodiversity. The criteria and thresholds provide a rigorous, data-driven approach that can be employed consistently across diverse taxonomic groups, ecosystems, and geographies. The requirements are designed to be comprehensive, obtaining various aspects of biodiversity importance from threatened species and ecosystems to unique biological processes and irreplaceable sites. The numerical thresholds provide clear guidelines while agreeing to apply expert judgment where data are restricted. This section briefly explains the criteria and thresholds for selecting KBAs [6, 7, 17] to ensure that only the most significant sites are prioritized for conservation efforts. Please go through this link for more detailed information on the criteria and thresholds: https://www.ibat-alliance.org/pdf/guidelines-for-kba-standard-v1.2.pdf.

1.4.1 Criteria for Selecting KBAs

(a) **Species-Based Criteria**

Species-based criteria emphasize sites supporting large populations of threatened species, restricted-range species, biome-restricted groups, and species that collect in large numbers. For instance, a site should support a significant proportion of a globally threatened species population, often at least 1%, to meet the criteria, helping safeguard critical habitats for species at risk of extinction [29, 30]. Species with limited geographic ranges are a crucial focus for conservation efforts. Sites that support a significant portion of these global species populations, typically around 5%, are given priority. Areas that host large congregations of animals, such as breeding grounds or migratory stopover sites, are also recognized. This approach ensures comprehensive protection across various habitats and life stages of different species.

(b) **Ecosystem-Based Criteria**

Ecosystem-focused criteria pinpoint areas that contain globally threatened ecosystems or maintain high ecological integrity with minimal human interference. These criteria emphasize protecting whole ecosystems, which is vital for biodiversity [31]. For example, a mangrove forest deemed globally threatened due to widespread degradation and limited range would be a priority KBA. Sites with high ecosystem integrity, where natural processes remain intact and minimal human impacts, are also crucial. Such areas serve as ecosystem health and resilience standards, providing essential services and supporting diverse species.

1.4.2 Thresholds for Selecting KBAs

Thresholds play a crucial role in determining whether a site qualifies as a KBA by providing quantitative benchmarks for assessment. These thresholds are applied differently depending on whether species-based or ecosystem-based criteria are being evaluated [32]. For species-based criteria, a standard threshold is the proportion of a species' global population that a site supports, with a typical requirement being at least 1% of the worldwide population [5]. This approach ensures that the most vital habitats for species preservation are identified and prioritized. In the case of ecosystem-based criteria, thresholds may focus on the percentage of a threatened area at a given site, aiming to safeguard significant portions of these vulnerable ecosystems. Furthermore, the evaluation process considers the representation of ecosystem types, considering their diversity, functionality, and integrity. This comprehensive approach helps identify and prioritize sites that best exemplify their respective ecosystem classifications, contributing to more effective conservation efforts.

1.4.3 Application and Considerations of Criteria and Thresholds

Identifying KBAs involves a rigorous data collection and analysis process, applying specific criteria and thresholds. This process draws on specialized approaches from various conservation initiatives. Important Plant Areas (IPAs), for instance, focus on botanical diversity, selecting sites based on the presence of a significant number of characteristic plant species. BirdLife International's Important Bird Areas (IBAs) program employs specific thresholds for congregatory species, such as designating areas that support at least 1% of a water bird species population globally. This approach ensures the protection of crucial sites for migratory birds. For example, a more tailored approach in Turkey was used, where sites supporting over 25% of bird species restricted to a particular terrestrial bioregion qualify as KBAs. This method highlights areas with high species richness and endemism. These diverse approaches collectively contribute to a comprehensive system for identifying and protecting the critical habitats for biodiversity conservation.

Data Requirements and Quality: Identifying KBAs relies heavily on collecting and analyzing high-quality data concerning species distribution, population sizes, and ecosystem conditions. This process employs various methods to gather comprehensive information. Remote sensing technologies offer valuable insights into ecosystem conditions and their changes over time, allowing for broader spatial and temporal analysis [33]. Additionally, citizen science programs have emerged as a significant contributor to biodiversity monitoring, engaging the public in data collection efforts and greatly expanding the scope and scale of available information. The integration of these diverse data sources is crucial for the accurate application of KBA criteria and thresholds.

Stakeholder Engagement: The identification and management of KBAs concern participating stakeholders, involving local communities, conservation organizations, and government agencies. Local communities' involvement ensures that conservation efforts are socially applicable and sustainable [34]. Conservation organizations offer expertise and resources for KBA identification and management, while government agencies play a crucial role in policy development and implementation. Effective stakeholder engagement ensures that conservation strategies are comprehensive and deliberate in the needs and knowledge of all parties included.

Climate Change Considerations: Climate change postures significant challenges for biodiversity conservation, changing species distributions, and ecosystem integrity [35–37]. When identifying KBAs, it is necessary to consider how climate change might impact these areas. Sites likely to endure resilience to climate change are prioritized, and adaptive management strategies are developed to address the changing environment. This proactive approach makes sure that KBAs continue to support biodiversity regardless of the impacts of climate change, preserving their conservation value over the long term.

1.5 Rationale for Selecting KBAs

The selection of KBAs for this study is motivated by different essential factors that support the primary goals of biodiversity conservation and sustainable development. The rationale for selecting KBAs is as follows:

(a) *Critical Role in Biodiversity Conservation*: KBAs are considered critically important for global biodiversity conservation. These areas provide habitats for many species, including rare, threatened, or endemic ones [5]. By focusing on KBAs, the study aims to ensure conservation efforts toward areas with the highest potential for preserving biodiversity.
(b) *Alignment with Sustainable Development Goals (SDGs)*: KBAs contribute directly to several SDGs, emphasizing the protection, restoration, and promotion of sustainable use of terrestrial ecosystems, sustainable forest management, combating desertification, and halting biodiversity loss [38]. By targeting KBAs, this study supports global efforts to achieve the SDG goals.
(c) *Vulnerability to Forest Loss and Fires*: KBAs are often located in regions highly vulnerable to deforestation and forest fires, posing significant threats to the ecological integrity of the KBA sites [39]. Understanding the impacts of these threats on KBAs can aid in developing targeted strategies to mitigate forest loss and manage fires efficiently.
(d) *Need for Monitoring and Management*: Given their ecological importance, there is urgent need to monitor KBAs regularly. Effective management and conservation policies need robust data on the status and trends of biodiversity within these areas [4]. This study focuses on developing and applying advanced monitoring technologies and policy frameworks to improve the protection of KBAs.

(e) *Technological Integration for Conservation*: The study aims to explore the technology–policy interface to combat forest loss and fires in KBAs. By leveraging advanced technologies such as remote sensing, geographic information systems (GIS), and data analytics [40], the study works toward developing modern solutions for monitoring and managing KBAs. This technological integration is necessary for improving the efficiency and effectiveness of conservation efforts.
(f) *Policy and Governance Implications*: The protection and management of KBAs need supportive policies and governance frameworks [4]. By focusing on KBAs, the study aims to identify and promote policy interventions that can improve the conservation of these critical areas. The findings will inform policy recommendations and contribute to developing robust governance frameworks for biodiversity conservation.
(g) *Improving Global and Local Conservation Initiatives*: Studying KBAs allows for aligning local conservation initiatives with global conservation priorities, ensuring that local actions contribute to global biodiversity targets and vice versa [32]. The findings of this study can thus support regional and international conservation efforts, encouraging collaboration and knowledge sharing.

In brief, the selection of KBAs for this study is based on their critical role in biodiversity conservation, alignment with SDGs, vulnerability to forest loss and fires, need for monitoring and management, representative nature, potential for technological integration, policy and governance implications, and ability to enhance global and local conservation initiatives. This approach ensures that the study addresses the most serious conservation challenges and contributes to sustainable biodiversity management.

1.6 Importance of Monitoring KBAs

Monitoring KBAs is vital for protecting biodiversity, advising conservation strategies, and achieving global sustainability goals. The monitoring ensures that conservation efforts are focused on the critical areas for biodiversity, extending their efficiency and impact. Monitoring KBA areas is crucial for various reasons:

(a) *Biodiversity Conservation*: Regular monitoring of KBAs allows conservationists to assess the effectiveness of current protection measures and implement adaptive management strategies when necessary [41]. It provides vital information on species populations, habitat conditions, and emerging threats, enabling targeted and efficient conservation actions.
(b) *Early Detection of Threats*: Monitoring allows for the early detection of threats such as deforestation, forest fires, and the encroachment of artificial lighting [42]. Early detection is necessary for timely interventions to mitigate these threats.

(c) *Evaluating Conservation Impact*: Monitoring provides data to assess the success of conservation policies and practices [4]. It helps in knowing the suitability of policies and practices, allowing for the optimization of conservation strategies.
(d) *Supporting SDGs*: KBAs are directly associated with several SDGs, specifically those related to life on land (SDG 15) and life below water (SDG 14) [43]. Monitoring KBAs contributes to achieving these goals by confirming that biodiversity is protected and sustainably managed.
(e) *Updating Policy and Decision-Making*: Data gathered from KBA monitoring updates policymakers and stakeholders, enabling evidence-based decision-making and essential for formulating policies and laws that support biodiversity conservation [4].
(f) *Participating and Empowering Local Communities*: Monitoring efforts frequently engage local communities, improving their participation and empowering them to participate in conservation actions [44]. This community participation is the solution to the long-term success of conservation initiatives.
(g) *Improving Global Biodiversity Databases*: Monitoring contributes to global biodiversity databases, providing beneficial information for scientists, conservationists, and policymakers globally. This data is vital for global biodiversity assessments and conservation planning.
(h) *Adapting to Climate Change*: Monitoring supports understanding the impacts of climate change on biodiversity within KBAs, allowing for the advancement of adaptation strategies to mitigate these impacts and improve ecosystem resilience.

1.7 Conclusion

KBAs are critical for global biodiversity conservation efforts. These sites of exceptional importance for preserving biodiversity play a vital role in achieving multiple sustainable development goals (SDGs) and contribute to broader ecological and social objectives. The global distribution of KBAs, with over 16,000 sites identified worldwide, highlights the diverse and widespread nature of critical habitats requiring protection. The varying sizes and distributions of KBAs across different regions underscore the need for tailored conservation strategies considering local ecological contexts and challenges. The standardized criteria and thresholds for KBA selection ensure a rigorous, data-driven approach to recognizing these vital areas. This systematic method allows for consistent application across various taxonomic groups, ecosystems, and geographies, enabling global conservation prioritization. The rationale for focusing on KBAs stems from their critical role in biodiversity conservation, alignment with SDGs, vulnerability to threats such as deforestation and fires, and potential for integrating advanced monitoring technologies. Conservation initiatives can maximize their impact and contribute significantly to global biodiversity targets by concentrating on these areas.

Monitoring KBAs facilitates the tracking of biodiversity health and status, allows for early detection of threats, provides data for measuring conservation impact, and supports evidence-based policy-making. Furthermore, monitoring efforts engage and empower local communities in conservation, contribute to global biodiversity databases, and aid in developing strategies for climate change adaptation. As we face increasing conservation challenges, identifying, protecting, and monitoring KBAs will become even more critical. Focusing on these high-biodiversity-value areas can substantially improve the preservation of Earth's ecological integrity for future generations. For the long-term success of KBA conservation efforts, it is essential to continue refining identification methods for monitoring techniques and strengthening the policy frameworks supporting their conservation. Integrating advanced technologies, fostering robust stakeholder engagement, and implementing adaptive management strategies will be vital to achieving these goals. By prioritizing KBAs and employing these strategies, we can make significant strides in preserving the planet's biodiversity for generations to come.

References

1. IUCN, *A Global Standard for the Identification of Key Biodiversity Areas (Version 1.0)* (2016)
2. D. Nigel, J.L. Boucher, A. Cuttelod et al., *Applications of Key Biodiversity Areas: End-User Consultations* (2014)
3. H. Costa, *Mozambique: Leading the Way on the Identification of Key Biodiversity Areas* (Wildlife Conservation Society, 2021). https://wildlifeconservationsociety.medium.com/mozambique-leading-the-way-on-the-identification-of-key-biodiversity-areas-efcfaa6c7d31
4. Bird Conservation International, Important Bird and Biodiversity Areas (IBAs): the development and characteristics of a global inventory of key sites for biodiversity. Bird Conserv. Int. **29**, 177–198 (2019). https://doi.org/10.1017/S0959270918000102
5. The KBA Partnership, *Guidelines on Business and KBAs: Managing Risk to Biodiversity* (2018). https://doi.org/10.2305/IUCN.CH.2018.05.en
6. IUCN, *Identification and Gap Analysis of Key Biodiversity Areas* (Gland, Switzerland, 2007)
7. IUCN, *Biodiversity for Business: A Guide to Using Knowledge Products Delivered Through IUCN* (Gland, Switzerland, 2014)
8. UNEP, *Conserve and Sustainably Use the Oceans, Seas and Marine Resources for Sustainable Development* (2014)
9. T. Cheskey, *Key Biodiversity Areas: What They Are and Why We Care* (Nature Canada, 2017). https://naturecanada.ca/news/blog/key-biodiversity-areas-what-they-are-and-why-we-care/
10. MBU, *Key Biodiversity Area Initiatives* (Marine Biodiversity Unit, 2024). https://sites.wp.odu.edu/GMSA/initiatives/kba/
11. DOPA, *DOPA Factsheet: Key Biodiversity Areas* (2023)
12. V. Jain, K.S. Rautela, M.K. Goyal, *Ecological Restoration: An Overview of Science and Policy Regime* (2023), pp. 1–27
13. S. Baidya, P. Chakraborty, S. Rakkasagi et al., *Pathways to Build Resilience Toward the Impact of Climate Change on the Indian Sunderban* (2023), pp. 307–333
14. P. Patle, P.K. Singh, S. Rakkasagi et al., *Application of Water Accounting Plus Framework for the Assessment of the Water Consumption Pattern and Food Security* (2023), pp. 257–269
15. S. Bhardwaj, P. Machiwar, C. Kant et al., *Analysis of Urbanization and Assessment of Its Impact on Groundwater and Land Use/Land Cover Using GIS Techniques: A Case Study of Bhopal and Gurugram District* (2023), pp. 219–255

References

16. CEPF, *What Is a Key Biodiversity Area? The Answer to That and All Your Other KBA-Related Questions* (Critical Ecosystem Partnership Fund, 2021). https://www.cepf.net/stories/what-key-biodiversity-area
17. P. Atukunda, W.B. Eide, K.R. Kardel et al., Unlocking the potential for achievement of the UN sustainable development goal 2—'zero hunger'—in Africa: targets, strategies, synergies and challenges. Food Nutr. Res. **65** (2021). https://doi.org/10.29219/fnr.v65.7686
18. M. Taka, L. Ahopelto, A. Fallon et al., The potential of water security in leveraging agenda 2030. One Earth **4**, 258–268 (2021). https://doi.org/10.1016/j.oneear.2021.01.007
19. R. Kumar, M.K. Goyal, R.Y. Surampalli, T.C. Zhang, River pollution in India: exploring regulatory and remedial paths. Clean Technol. Environ. Policy (2024). https://doi.org/10.1007/s10098-024-02763-9
20. V. Gupta, S. Rakkasagi, S. Rajpoot et al., Spatiotemporal analysis of Imja Lake to estimate the downstream flood hazard using the SHIVEK approach. Acta Geophys. (2023). https://doi.org/10.1007/s11600-023-01124-2
21. P. Schröder, A.S. Antonarakis, J. Brauer et al., SDG 12: responsible consumption and production—potential benefits and impacts on forests and livelihoods, in *Sustainable Development Goals: Their Impacts on Forests and People* (Cambridge University Press, 2019), pp. 386–418
22. S. Rakkasagi, M.K. Goyal, S. Jha, Evaluating the future risk of coastal Ramsar wetlands in India to extreme rainfalls using fuzzy logic. J. Hydrol. **632**, 130869 (2024). https://doi.org/10.1016/j.jhydrol.2024.130869
23. Y. Zhang, Y. Li, J. Liu, Global decadal assessment of life below water and on land. iScience **26**, 106420 (2023). https://doi.org/10.1016/j.isci.2023.106420
24. N. Kumar, M.K. Goyal, Projected changes in monsoonal compound dry-hot extremes in India. Atmos. Res. **310**, 107605 (2024). https://doi.org/10.1016/j.atmosres.2024.107605
25. M.K. Goyal, S. Rakkasagi, S. Shaga et al., Spatiotemporal-based automated inundation mapping of Ramsar wetlands using Google Earth Engine. Sci. Rep. **13**, 17324 (2023). https://doi.org/10.1038/s41598-023-43910-4
26. K.S. Rautela, S. Singh, M.K. Goyal, Aerosol atmospheric rivers: patterns, impacts, and societal insights. Environ. Sci. Pollut. Res. (2024). https://doi.org/10.1007/s11356-024-34625-8
27. K.S. Rautela, S. Singh, M.K. Goyal, Resilience to air pollution: a novel approach for detecting and predicting aerosol atmospheric rivers within earth system boundaries. Earth Syst. Environ. (2024). https://doi.org/10.1007/s41748-024-00421-0
28. S. Rakkasagi, M.K. Goyal, Assessing risk levels of the extreme rainfalls in Ramsar wetlands of India using fuzzy logic, in *Chapman Conference on Remote Sensing of the Water Cycle* (AGU, 2024)
29. M. Kumar Goyal, V. Poonia, V. Jain, Three decadal urban drought variability risk assessment for Indian smart cities. J. Hydrol. **625**, 130056 (2023). https://doi.org/10.1016/j.jhydrol.2023.130056
30. V. Poonia, M. Kumar Goyal, S. Jha, S. Dubey, Terrestrial ecosystem response to flash droughts over India. J. Hydrol. **605**, 127402 (2022). https://doi.org/10.1016/j.jhydrol.2021.127402
31. J.P. Rodríguez, K.M. Rodríguez-Clark, J.E.M. Baillie et al., Establishing IUCN red list criteria for threatened ecosystems. Conserv. Biol. **25**, 21–29 (2011). https://doi.org/10.1111/j.1523-1739.2010.01598.x
32. IUCN, *Guidelines for Using a Global Standard for the Identification of Key Biodiversity Areas: Version 1.2* (IUCN, International Union for Conservation of Nature, Gland, Switzerland, 2022)
33. A.E. Beresford, P.F. Donald, G.M. Buchanan, Repeatable and standardised monitoring of threats to Key Biodiversity Areas in Africa using Google Earth Engine. Ecol. Indic. **109**, 105763 (2020). https://doi.org/10.1016/j.ecolind.2019.105763
34. J. Maxwell, S. Allen, T. Brooks et al., Engaging end-users to inform the development of the global standard for the identification of Key Biodiversity Areas. Environ Sci. Policy **89**, 273–282 (2018). https://doi.org/10.1016/j.envsci.2018.07.019
35. M.K. Goyal, A.K. Gupta, J. Das et al., Heatwave magnitude impact over Indian cities: CMIP 6 projections. Theor. Appl. Climatol. **154**, 959–971 (2023). https://doi.org/10.1007/s00704-023-04599-7

36. S. Rakkasagi, V. Poonia, M.K. Goyal, Flash drought as a new climate threat: drought indices, insights from a study in India and implications for future research. J. Water Clim. Change (2023). https://doi.org/10.2166/wcc.2023.347
37. M.K. Goyal, A.K. Gupta, S. Jha et al., Climate change impact on precipitation extremes over Indian cities: non-stationary analysis. Technol. Forecast. Soc. Change **180**, 121685 (2022). https://doi.org/10.1016/j.techfore.2022.121685
38. P.F. Donald, G.M. Buchanan, A. Balmford et al., The prevalence, characteristics and effectiveness of Aichi Target 11's "other effective area-based conservation measures" (OECMs) in Key Biodiversity Areas. Conserv. Lett. **12** (2019). https://doi.org/10.1111/conl.12659
39. IUCN, *Deforestation and Forest Degradation* (IUCN, 2021)
40. S.R. Subramoniam, S. Ravindranath, S. Rakkasagi, H. Ram, *Water Resource Management Studies at Micro Level Using Geospatial Technologies* (2022), pp. 49–74
41. X. Dong, J. Gong, W. Zhang et al., Importance of including Key Biodiversity Areas in China's conservation area-based network. Biol. Conserv. **296**, 110676 (2024). https://doi.org/10.1016/j.biocon.2024.110676
42. Z. Waliczky, L.D.C. Fishpool, S.H.M. Butchart et al., Important Bird and Biodiversity Areas (IBAs): their impact on conservation policy, advocacy and action. Bird Conserv. Int. **29**, 199–215 (2019). https://doi.org/10.1017/S0959270918000175
43. NITI Aayog, *National Consultation on SDGs—Sustaining Life: Integrating Biodiversity Concerns, Ecosystems Values and Climate Resilience in India's Planning Process Focus on SDG 13, 14 and 15* (2017)
44. J. Das, M.K. Goyal, N.V. Umamahesh, Water harvesting, climate change, and variability, in *Handbook of Water Harvesting and Conservation* (Wiley, 2021), pp. 427–446

Chapter 2
Key Biodiversity Areas and Forest Cover Loss

2.1 Introduction

In the face of accelerating global ecological change, conserving biodiversity has appeared as one of the most critical challenges of the present time [1]. The Earth's forests, home to an estimated 80% of terrestrial biodiversity, stand at the forefront of this conservation combat [2, 3]. These complex ecosystems not only harbor a substantial group of plant and animal species but also play crucial roles in climate adaptation, water cycle management, and the provision of essential ecosystem services [4–6]. However, the persistent expansion of anthropogenic activities has led to exceptional deforestation and forest degradation rates, threatening the fabric of the biodiversity hotspots. Key Biodiversity Areas (KBAs) have appeared as a robust framework in the global effort to identify and safeguard the most critical sites for biodiversity conservation. Described as "sites contributing significantly to the global persistence of biodiversity" [7], KBAs represent a standardized approach to identifying areas of particular importance for biodiversity conservation. This framework was developed by collaborating with international conservation organizations to provide a scientifically rigorous method for identifying sites crucial for preserving species populations, ecosystems, and ecological processes [8, 9].

The importance of KBAs in forest conservation cannot be overstated. Many KBAs encompass forested areas that are not only rich in biodiversity but also under significant threat from anthropogenic activities. The connection of KBAs with forest ecosystems presents severe attention to conservation efforts, as these areas often represent the last strongholds for many endangered and endemic species [10, 11]. Therefore, understanding the dynamics of forest loss within KBAs is crucial for developing effective conservation strategies and policies. Forest loss, driven by a complex interplay of factors, including agricultural expansion, logging, infrastructure development, and climate change, directly threatens the integrity of KBAs [12–15]. The rapid pace of deforestation in many parts of the world has overtaken the capacity of natural systems to regenerate, leading to permanent loss of habitat and biodiversity.

Forest loss has been particularly severe in tropical regions, which host most of the world's biodiversity. Recent studies have shown that tropical forests continue to be lost at alarming rates, with an estimated 12 million hectares of tropical tree cover loss in 2020 alone [16]. The impacts of forest loss on biodiversity are profound and multifaceted. Beyond the abrupt loss of habitat, deforestation destroys remaining forest areas, disturbing ecological connectivity and altering species distributions [17, 18]. This destruction can have cascading effects on ecosystem functions, including pollination, seed dispersal, and nutrient cycling [19]. Moreover, the loss of forest cover can worsen climate change by reducing carbon sequestration capacity and changing local and regional climate patterns [20–22].

In recent years, advancements in remote sensing technologies and data analysis techniques have revolutionized our ability to monitor and quantify forest loss at various scales [23]. Global forest cover change datasets, such as those produced by Hansen et al. [14] and updated annually, provide exceptional insights into the patterns and trends of deforestation worldwide. Recent developments, including radar and LiDAR data integration, have further improved our ability to assess forest structure and biomass [24]. When overlaid with KBA boundaries, these datasets offer a powerful tool to determine the state of forest cover within these critical areas and identify hotspots of forest loss that require urgent conservation attention. The technology–policy interface is crucial in translating scientific knowledge about forest loss in KBAs into effective conservation action. Remote sensing data and advanced analytical tools can inform policy decisions by providing timely and accurate information on forest cover changes [25, 26]. However, bridging the gap between scientific evidence and policy implementation remains a significant challenge. Effective conservation strategies must navigate complex socioeconomic landscapes, balancing the imperative of biodiversity protection with the needs of local communities and national development goals [27, 28]. This study aims to understand the spatiotemporal patterns of forest loss within KBAs, focusing on India. By integrating spatial analysis of forest cover change with an assessment of KBA effectiveness, we seek to provide insights to inform more targeted and effective conservation strategies. Furthermore, we explore the potential of emerging technologies and policy approaches to strengthen the protection of these critical biodiversity areas.

2.2 Impacts of Deforestation on Biodiversity

Deforestation, the large-scale removal of forest ecosystems, denotes one of the most significant threats to global biodiversity. As repositories of biological richness, forests play a crucial role in maintaining the Earth's ecological balance. The impacts of deforestation on biodiversity are profound, multifaceted, and often permanent, influencing ecosystems at local, regional, and global scales. Deforestation's most direct and abrupt impact on biodiversity is the loss and fragmentation of habitats. Forests provide complex, three-dimensional living spaces for many species, from the forest surface to the canopy. When forests are cleared, these habitats are destroyed, leading

2.2 Impacts of Deforestation on Biodiversity

to abrupt declines in population sizes and, in many cases, local extinctions [17]. Habitat fragmentation, a consequence of partial deforestation, can be equally detrimental as continuous forests are broken into smaller patches, edge effects increase, altering microclimates and exposing forest interiors to external pressures. This fragmentation can disturb species interactions, limit gene flow between populations, and reduce the ability of species to move in response to climate change [18]. The loss of forest habitats directly contributes to species extinctions. Many forest-dependent species, particularly those with small ranges or specific habitat requirements, face increased extinction risk due to deforestation. The IUCN Red List of Threatened Species identifies habitat loss, predominantly due to deforestation, as the major threat to 85% of all species described on the Red List [29]. Tropical forests harbor the highest levels of terrestrial biodiversity and are particularly vulnerable. A study by Alroy [30] projected that up to 57% of tropical forest species are threatened with extinction due to deforestation. This loss extends beyond the immediately apparent fauna and flora; it includes countless microorganisms and undiscovered species that may disappear before they are even known to science.

Deforestation disturbs the complex web of ecological processes that support forest ecosystems. Many forest plants depend on animal pollinators, which may decline or vanish with deforestation, affecting plant reproduction and genetic diversity [31]. The loss of frugivorous animals due to deforestation can disrupt seed dispersal mechanisms, altering forest restoration patterns and composition. Deforestation alters nutrient cycles by removing biomass, increasing erosion, and changing soil microbial communities. The loss of top predators or key prey species can trigger trophic cascades, primarily altering ecosystem structure and function [32]. Deforestation compromises these services, with cascading effects on biodiversity. Forests play a vital role in global climate regulation, and deforestation contributes to climate change by releasing carbon and decreasing carbon sequestration capacity [22]. Climate change, in turn, causes a significant threat to global biodiversity [33]. Forests influence local and regional precipitation patterns through evapotranspiration [34]. Deforestation can lead to reduced rainfall and increased drought [35, 36], affecting biodiversity in forest and non-forest ecosystems [18]. Forest removal increases soil erosion, leading to sedimentation in aquatic ecosystems and loss of soil biodiversity [37].

The impacts of deforestation on biodiversity often involve complex, cascading effects that can take years or decades to manifest fully. This phenomenon, known as "extinction debt," refers to the time lag between habitat loss and species extinctions [38]. Even after deforestation, species may continue for some time before eventually dying out, meaning that the full biodiversity impacts of current deforestation may not be realized for years. Some generalist species may adapt to or benefit from forest disturbance, potentially leading to biotic homogenization as forest specialists are replaced by more common, widespread species [39]. This shift can result in a net loss of biodiversity even if the total number of species remains relatively stable. Deforestation not only affects species diversity but also genetic diversity within species. As populations become smaller and more isolated due to habitat loss and fragmentation, they face increased genetic drift and inbreeding, potentially reducing their adaptive capacity and long-term viability [40].

2.3 Datasets and Methodology for Forest Cover Losses

2.3.1 Data Sources and Preprocessing

The study used spatial data on KBAs in India obtained from the World Database of Key Biodiversity Areas maintained by BirdLife International on behalf of the KBA Partnership. This dataset contains polygons representing the boundaries of identified KBAs across India. To analyze forest cover loss, we employed the Hansen Global Forest Change dataset [14], derived from Landsat imagery, which provides global tree cover and annual forest loss information at a 30-m resolution. Specifically, we used the tree cover in the year 2000 (treecover2000) layer, representing the percentage of tree cover per pixel for the baseline year 2000, and the year of gross forest cover loss event (lossyear) layer, indicating the year of detected forest loss for each pixel between 2001 and 2020. The Hansen dataset was accessed and processed through Google Earth Engine (GEE), a cloud-based Earth science data analysis platform.

2.3.2 Google Earth Engine Implementation

The KBA shapefile was imported into Google Earth Engine as a "FeatureCollection," with each KBA polygon assigned a unique identifier to facilitate individual site analysis. Preprocessing steps included reprojection to match the coordinate reference system of the forest cover data, filtering to include only KBAs within India's administrative boundaries, and removing any invalid geometries or duplicate features. The analysis of forest cover loss within KBAs was implemented using a workflow in Google Earth Engine that included several vital steps. First, a tree cover threshold was applied to the treecover2000 layer to create a binary forest/non-forest mask, with pixels having a tree cover percentage greater than or equal to 50% classified as forest. This threshold aligns with the FAO definition of forest and helps exclude areas with sparse tree cover. The baseline forest area for each KBA was then calculated by summing the area of all pixels classified as forest within the KBA boundary in 2000. Annual forest loss computation utilized the lossyear layer to create annual forest loss maps for each year from 2001 to 2020. For each year, pixels indicating loss were identified, and their areas were summed within each KBA. Relative forest loss was calculated by expressing annual forest loss as a percentage of the baseline forest area in 2000 for each KBA, allowing for comparison between KBAs of different sizes. Finally, cumulative forest loss from 2001 to 2020 was calculated for each KBA by summing the annual loss values.

2.3 Datasets and Methodology for Forest Cover Losses

2.3.3 Statistical Analysis

To assess the trends in forest cover loss over time for each KBA, we employed the Mann–Kendall test, a non-parametric test for monotonic trends in time series data [41]. For each KBA, the annual relative forest loss percentages from 2001 to 2020 were used as the time series. The null hypothesis of no trend was tested against the alternative hypothesis of an increasing or decreasing trend, with the significance level set at $\alpha = 0.05$. Sen's slope estimator was calculated to quantify the magnitude of the trend. The Mann–Kendall test was chosen for its robustness to outliers and ability to handle non-normally distributed data, which is common in ecological time series [26]. Based on the results of the Mann–Kendall test, each KBA was classified into one of four categories: Significant positive trend (p-value < 0.05 and positive Sen's slope), positive trend (positive Sen's slope but p-value $\geq$ 0.05), negative trend (negative Sen's slope but p-value $\geq$ 0.05), and significant negative trend (p-value < 0.05 and negative Sen's slope). This classification allows for a nuanced understanding of forest loss patterns, distinguishing between statistically significant trends and directional tendencies.

2.3.4 Spatial and Temporal Analysis

KBAs were grouped by Indian states and union territories to provide insights into spatio-temporal patterns. We calculated the proportion of KBAs falling into each trend category for each state, computed mean annual forest loss rates, and summarized cumulative forest loss from 2001 to 2020. This state-level analysis allows for identifying regional hotspots of forest loss and prosperous conservation areas. The annual forest loss rates were aggregated across all KBAs for each year from 2001 to 2020 to examine temporal patterns in forest loss. A time series plot was created to visualize overall trends, and change-point analysis was performed to identify significant shifts in forest loss rates over the study period. To identify KBAs experiencing the most severe forest loss, we ranked KBAs based on their cumulative forest loss percentage from 2001 to 2020. The top 10% of KBAs with the highest loss were identified as hotspots, and spatial clustering of hotspots was assessed.

2.3.5 Validation and Uncertainty Assessment

To validate the forest loss estimates derived from the Hansen dataset, a stratified random sample of 100 KBAs was selected. High-resolution satellite imagery (i.e., from Google Earth) was manually interpreted for these sample sites. Forest loss estimates from manual interpretation were compared with those derived from the Hansen dataset, and overall accuracy, user's accuracy, and producer's accuracy were calculated. Confidence intervals for annual and cumulative forest loss were calculated using bootstrapping techniques to quantify uncertainty in the forest loss estimates. The sensitivity of results to the tree cover threshold (50%) was assessed by repeating the analysis with 30 and 70% thresholds. Potential sources of error, including geolocation inaccuracies, mixed pixels, and seasonal variability, were discussed qualitatively.

2.3.6 Limitations and Considerations

Several limitations and considerations of the methodology were acknowledged. A 50% tree cover threshold may not capture all forest types, particularly in arid or semi-arid regions where natural forests may have lower canopy cover. The Hansen dataset does not distinguish between natural forests and plantations, potentially overestimating forest loss in areas of plantation rotation. The 30-m resolution of the Hansen dataset may not capture small-scale forest disturbances or degradation. Persistent cloud cover in some regions may lead to data gaps or inaccuracies in forest loss detection. Any changes to KBA boundaries during the study period were not accounted for, as the most recent KBA shapefile was used. The analysis is also limited by the temporal coverage of the Hansen dataset, which currently extends only to 2020.

2.4 Results and Discussion of Forest Loss in Indian KBAs

2.4.1 Overall Trend Analysis

Our analysis of forest cover loss trends in KBAs across India reveals a complex picture with both encouraging signs of conservation success and concerning areas of ongoing deforestation. Using Google Earth Engine to analyze the Hansen Global Forest Change dataset from 2001 to 2020, we identified several critical patterns in forest loss across KBAs in India (Fig. 2.1). The results show a mixed picture of forest cover change in KBAs in India from 2001 to 2020. About 32.05% of KBAs showed a statistically significant negative trend in forest loss, indicating a reduction in deforestation rates over time in these areas, encouraging signs that conservation efforts may positively impact nearly a third of KBAs in India. Additionally, 17.54% of KBAs showed a negative trend in forest loss that was not statistically significant. While inconclusive, this suggests that deforestation may also be slowing in these areas. However, 13.07% of KBAs showed a positive trend in forest loss that was not statistically significant, indicating ongoing deforestation pressure in these areas, though the trend is not definitive. More concerningly, 10.36% of KBAs showed a statistically significant positive trend in forest loss, pointing to accelerating deforestation in these areas of high biodiversity importance, highlighting areas needing

2.4 Results and Discussion of Forest Loss in Indian KBAs

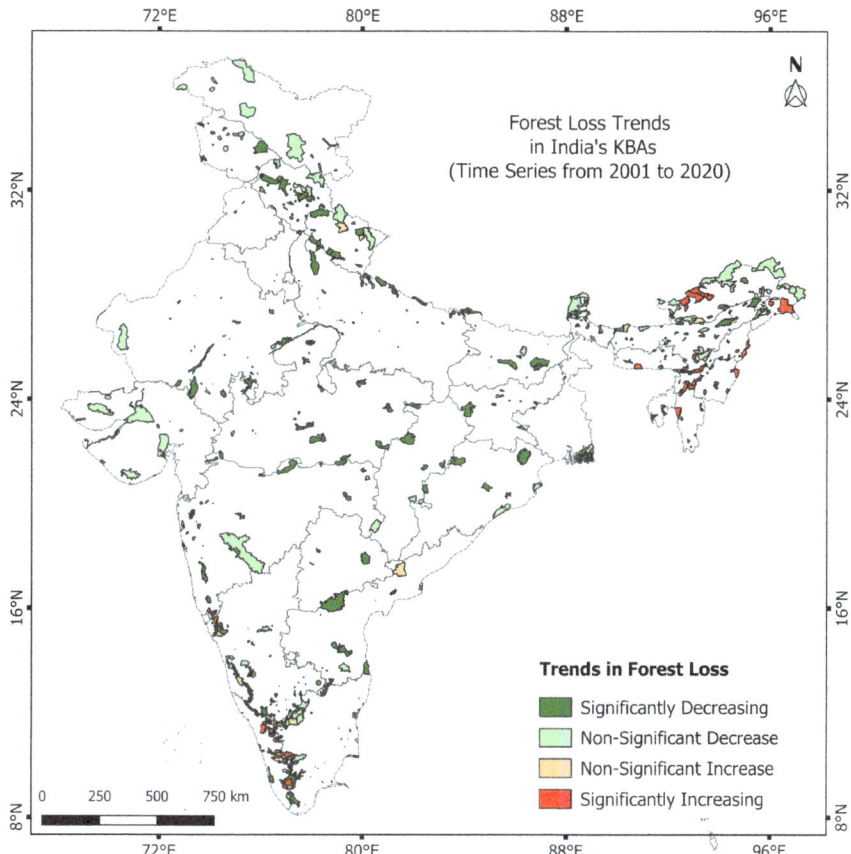

Fig. 2.1 Map visually represents the spatial distribution of forest loss trends across KBAs in India over the two-decade period (from 2001 to 2020), highlighting areas of concern and improvement in forest conservation efforts

urgent conservation intervention. The remaining 26.98% of KBAs showed no clear trend in forest loss rates over the study period. These overall figures are somewhat optimistic, with nearly 50% of KBAs showing negative trends in forest loss (both significant and non-significant) compared to only about 23% showing positive trends. However, the fact that over 10% of these critical biodiversity areas are experiencing significantly increasing rates of forest loss is cause for serious concern.

2.4.2 Regional Patterns

Our analysis revealed some notable regional patterns in forest cover change across KBAs in India. The Northeastern states, including Arunachal Pradesh, Assam,

Manipur, Meghalaya, and Nagaland, showed some of the most concerning trends in forest loss (Fig. 2.2). In Arunachal Pradesh, 59.47% of KBAs showed positive trends in forest loss, with three sites showing statistically significant increases, aligning with other studies that have identified high deforestation rates in this biodiversity-rich state. Manipur showed the most severe forest loss trends, with 57.14% of KBAs exhibiting statistically significant positive trends in deforestation, pointing to a major conservation crisis in the state's protected areas. Meghalaya also showed worrying patterns, with 62.5% of KBAs experiencing significant positive trends in forest loss. Nagaland similarly had a high proportion of KBAs with significant tree loss trends.

These patterns in the Northeast are particularly troubling given the region's exceptional biodiversity and high concentration of endemic species. The Eastern Himalayas biodiversity hotspot, which encompasses much of Northeast India, is facing severe threats from deforestation. Our results suggest that even designated regional KBAs are not immune to these pressures. Several factors may be driving these high forest loss rates in Northeastern KBAs, including shifting cultivation practices, illegal logging, infrastructure development, expansion of commercial plantations, and weak enforcement of forest protection laws in some remote areas. In contrast to the Northeast, several states in Western and Central India showed more positive trends. In

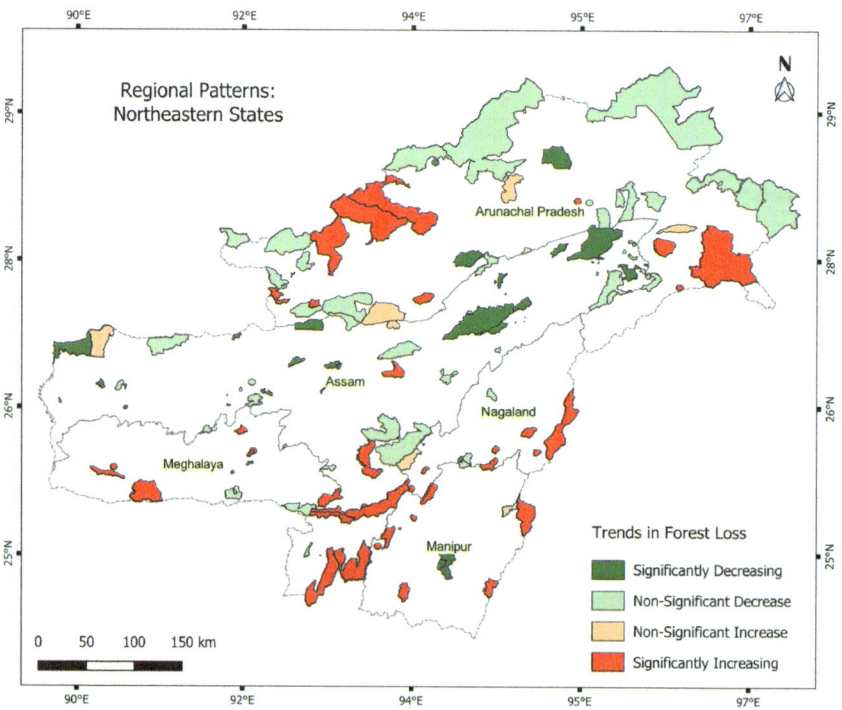

Fig. 2.2 Map visually represents the spatial distribution of forest loss trends across KBAs in the Northeastern states of India over the two-decade period (from 2001 to 2020)

2.4 Results and Discussion of Forest Loss in Indian KBAs

Gujarat, only one KBA showed a significantly increasing trend in forest loss. The majority exhibited stable or declining deforestation rates. Madhya Pradesh's KBAs uniformly showed negative trends in forest loss, suggesting successful conservation interventions in the state's protected areas. Maharashtra similarly showed predominantly negative trends in forest loss across its KBAs. All of Chhattisgarh's KBAs showed negative trends in forest loss, with 77.78% exhibiting statistically significant declines in deforestation rates, indicating a notable conservation success story. These more positive trends in central Indian states may reflect more vigorous enforcement of forest protection laws, successful community-based conservation initiatives, and lower demographic pressures than other regions. The significant improvements in Chhattisgarh are noteworthy and warrant further study to identify replicable conservation strategies.

The Southern states showed mixed results. Karnataka's KBAs primarily showed negative trends in forest loss, aligning with the state's reputation for progressive forest management (Fig. 2.3). Kerala exhibited positive and negative trends, with no clear overall pattern emerging. Tamil Nadu showed varied trends in forest loss across its KBAs, precluding any general conclusions for the state. These diverse trends in the South likely reflect the heterogeneous landscapes and varied anthropogenic pressures across the region's KBAs. Site-specific analysis would be needed to unpack the drivers of forest loss in different Southern KBAs.

The Himalayan states generally showed encouraging trends. In Himachal Pradesh, all KBAs exhibited negative trends in forest loss for 2001–2020, with none showing significant increases in deforestation. Uttarakhand similarly showed predominantly negative trends in forest loss across its KBAs. Jammu & Kashmir's KBAs uniformly displayed negative trends in forest loss. These results suggest that conservation efforts in Himalayan KBAs have mainly been successful in curbing deforestation over the past two decades, given the Himalayas' importance as a biodiversity hotspot and the sensitivity of montane ecosystems to disturbance. Factors contributing to these positive trends may include strict protected area management, inaccessibility of many sites, and lower population pressures compared to lowland areas.

2.4.3 *Implications and Discussion*

The heterogeneous trends in forest cover change across KBAs in India have several important implications for biodiversity conservation in the country:

(a) *Targeted Interventions Needed*: The entire regional differences in forest loss trends highlight the need for targeted, context-specific conservation interventions. Given the diverse drivers of deforestation across India's varied landscapes, one-size-fits-all approaches are unlikely to be effective.

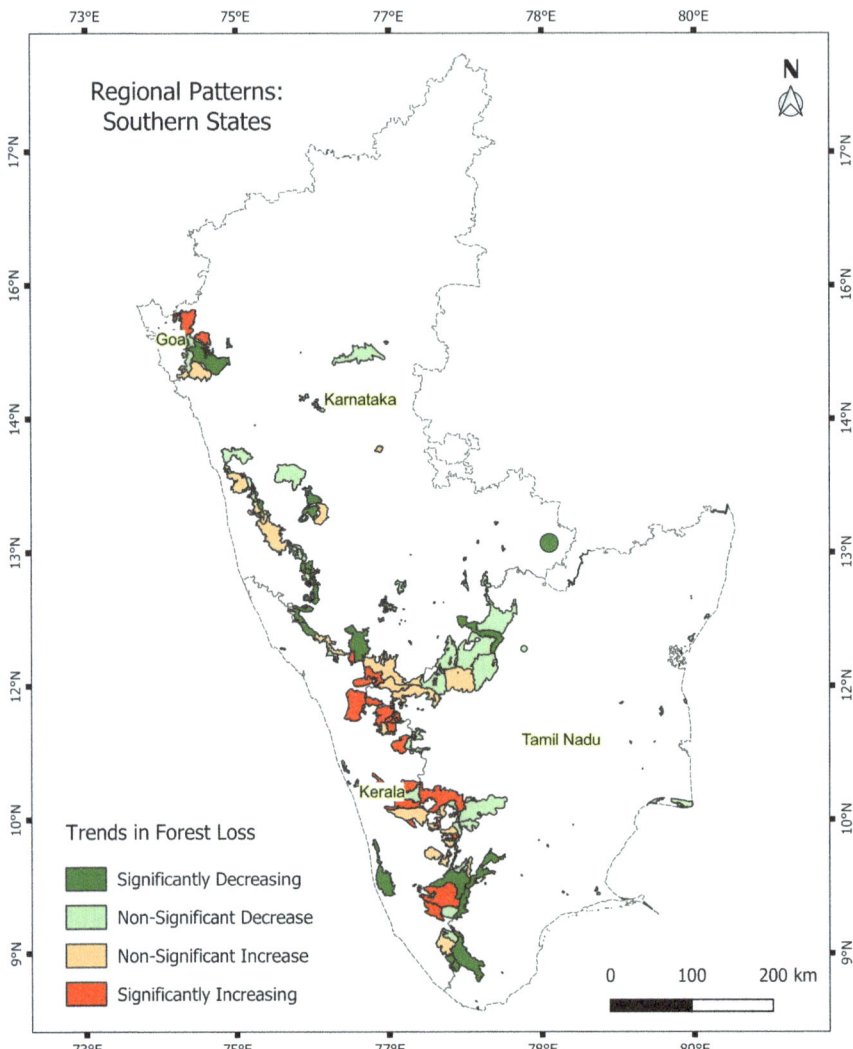

Fig. 2.3 Map visually represents the spatial distribution of forest loss trends across KBAs in the Southern states of India over the two-decade period (from 2001 to 2020)

(b) *Northeast India Crisis*: The alarming rates of forest loss in Northeastern KBAs demand urgent attention. Strengthened enforcement, community-based conservation initiatives, and sustainable livelihood programs may all be needed to address the complex drivers of deforestation in the region.

(c) *Learning from Successes*: The states in the central India (i.e., Chhattisgarh and Madhya Pradesh) demonstrate that reducing deforestation rates in KBAs is

2.4 Results and Discussion of Forest Loss in Indian KBAs

possible. Further research should investigate the policy measures, management practices, and local contexts that have enabled these conservation successes.

(d) *Importance of Long-Term Monitoring*: Our analysis demonstrates the value of consistent, long-term monitoring of forest cover change using remote sensing. Such data is crucial for identifying trends, assessing conservation effectiveness, and guiding adaptive management of protected areas.

(e) *Limitations of Forest Cover as a Metric*: While forest cover loss is an essential indicator of ecosystem health, it does not capture all relevant changes in biodiversity and habitat quality. For instance, our analysis does not account for forest degradation that falls short of complete canopy loss, nor does it distinguish between natural forests and plantations. Complementary indicators and ground-based monitoring remain essential.

(f) *Climate Change Considerations*: As climate change alters temperature and precipitation patterns across India [42–45], some shifts in forest cover may reflect vegetation responses to changing environmental conditions rather than direct anthropogenic impacts, underscoring the need for climate-adaptive conservation strategies in KBAs.

(g) *Policy Implications*: Our results suggest that India's forest conservation policies have had mixed success in protecting KBAs. While some regions show clear improvements, the ongoing loss of forest cover in many critical biodiversity areas indicates that more robust protective measures may be needed. The protective measures could include expanded protected area coverage, improved enforcement of existing regulations, and increased support for community-based conservation initiatives.

(h) *Transboundary Considerations*: Many of KBAs in India are part of larger transboundary ecosystems, particularly in the Northeast and Himalayas. Effective long-term conservation will require international cooperation in monitoring and managing these shared landscapes.

2.4.4 Limitations and Future Research Directions

While our analysis provides valuable insights into forest cover trends in KBAs in India, several limitations should be noted. The 30-m resolution of the Hansen dataset, while suitable for national-scale analysis, may miss fine-scale forest changes that could be ecologically significant in some KBAs. Our analysis focuses solely on forest cover loss and does not account for degradation or changes in forest quality that fall short of complete canopy removal. The Hansen dataset does not differentiate between causes of forest loss (e.g., logging, fire, storm damage). A more detailed analysis would be needed to attribute observed changes to specific drivers. Our analysis covers 2001–2020, but forest loss patterns may have shifted in very recent years due to policy changes or emerging pressures. Finally, we treated KBA boundaries as fixed, but these may be adjusted over time. Future analyses could incorporate temporal changes in KBA delineations.

Future research directions that could build on this work include integrating multiple remote sensing products to gain a more comprehensive picture of forest health and biodiversity in KBAs, conducting fine-scale analyses of select KBAs to understand local drivers of forest loss better and identify site-specific conservation needs, investigating the effectiveness of different protected area management regimes in maintaining forest cover within KBAs, assessing how forest loss patterns in KBAs compare to surrounding landscapes to evaluate the additionality of KBA designations, exploring the relationships between forest cover trends and population changes of key species within KBAs, and projecting future forest cover scenarios for KBAs under different climate change and land-use trajectories.

2.5 Technology–Policy Interface to Reduce Deforestation

Integrating advanced technologies with effective policy measures presents a promising approach to combating deforestation in KBAs in India. This interface can enhance monitoring capabilities, improve enforcement, and inform evidence-based policy-making (Fig. 2.4). Advanced technologies and community involvement revolutionize forest conservation. Remote sensing and satellite imagery provide unprecedented detail in forest monitoring, enabling the detection of subtle changes and rapid response to deforestation [46]. Multispectral analysis offers insights into forest health, allowing early intervention before significant damage occurs. By combining these technological tools with data analytics, conservationists can develop more effective strategies to protect vital habitats. Additionally, engaging local communities in conservation efforts ensures that strategies are scientifically sound, practical, and culturally sensitive. This integrated approach, merging cutting-edge technology with grassroots participation, represents the future of forest preservation and sustainable ecosystem management.

The power of data analytics and machine learning offers tremendous potential for predicting areas at high risk of deforestation, enabling proactive conservation measures [47]. By leveraging big data analytics, we can identify and quantify the key drivers of deforestation across different regions, providing crucial insights to inform targeted policy interventions. This technological approach allows policymakers to refine their strategies and allocate resources more efficiently by analyzing the effectiveness of various conservation policies and interventions using historical satellite data and advanced statistical methods. These technological insights can be directly applied to policy development and implementation. For instance, region-specific conservation policies can be designed to address the unique challenges driving deforestation in different areas, such as the distinct pressures faced in Northeast India compared to Western states. Adaptive management strategies for KBAs can be developed based on regular monitoring and analysis of forest cover trends, ensuring that conservation efforts remain effective as pressures evolve regularly. Additionally, policies that reward states or local communities for successful forest conservation, as measured by satellite-based monitoring, can create positive

2.5 Technology–Policy Interface to Reduce Deforestation

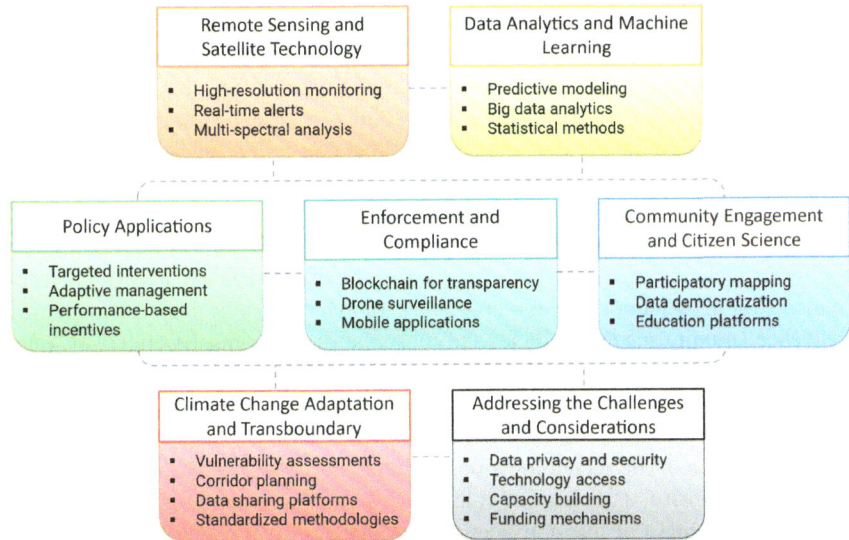

Fig. 2.4 Diagram outlines different technological tools and approaches that can be integrated to support the technology–policy interface to reduce deforestation, from data analysis to policy implementation in KBAs

incentives for protection. Technology can also significantly enhance enforcement and compliance efforts. Blockchain technology can be implemented to track timber products and combat illegal logging, improving supply chain transparency [48]. Drones can be deployed for targeted monitoring of high-risk areas or to investigate deforestation alerts in remote KBAs [49]. Furthermore, mobile applications can empower forest guards and local communities to report and document illegal activities in real-time, creating a more responsive and engaged protection network.

Community engagement and citizen science initiatives are vital in the technology–policy interface. Using user-friendly mobile technologies, participatory mapping can engage local communities in monitoring forest resources and changes. Making forest cover data and analysis tools publicly accessible empowers NGOs, researchers, and citizen scientists to contribute meaningfully to conservation efforts. Online platforms can be developed to educate stakeholders about the importance of KBAs and sustainable forest management practices, fostering broader understanding and support for conservation initiatives [50]. As climate change continues to impact forest ecosystems, the technology–policy interface must address these evolving challenges. Climate models and forest cover data can be used to assess the vulnerability of KBAs to climate change impacts, informing adaptive management strategies [12]. Spatial analysis tools can help identify and protect forest corridors that allow species migration in response to changing climates, ensuring the long-term viability of biodiversity in these areas. Given the transboundary nature of many ecosystems, especially in the Northeast and Himalayan regions, international cooperation is essential. Establishing data-sharing platforms can facilitate the monitoring and management of transboundary KBAs. Developing standardized methodologies for forest monitoring across countries can enable comparative analyses and joint conservation efforts, fostering a more cohesive approach to protecting shared ecosystems.

However, implementing this technology–policy interface is not without challenges. Data privacy issues must be carefully balanced with the need for detailed monitoring. The digital divide must be addressed to ensure that all relevant stakeholders have access to necessary technologies and data [51]. Significant investments in capacity building are needed to train local personnel in using and interpreting forest monitoring technologies. Innovative funding mechanisms, such as carbon markets or payments for ecosystem services, should be explored to support the implementation of advanced monitoring technologies and conservation programs. By effectively integrating these technological tools with robust policy frameworks, India can enhance its ability to protect and manage its KBAs. This approach can help curb deforestation, preserve critical habitats, and ensure the persistence of India's rich biodiversity for future generations. The technology–policy interface provides a pathway for more informed, responsive, and effective conservation strategies, addressing the complex and varied challenges of forest protection across diverse landscapes in India.

2.6 Conclusion

The comprehensive analysis of forest cover loss in KBAs in India from 2001 to 2020 presents a nuanced picture of conservation challenges and successes across the country. Researchers utilized advanced remote sensing techniques and the Hansen Global Forest Change dataset to uncover significant regional variations in deforestation trends, highlighting areas of concern and notable conservation achievements. The study's findings offer a cautiously optimistic outlook, with nearly half of KBAs in India demonstrating negative trends in forest loss rates, suggesting that conservation efforts have been practical in many areas, particularly in states like Chhattisgarh and Madhya Pradesh. However, the research also identifies critical hotspots of accelerating deforestation, especially in the biodiversity-rich Northeastern states, where urgent intervention is needed to protect vulnerable ecosystems and endemic species.

These results underscore the importance of tailored, context-specific conservation strategies. The stark regional differences in forest loss trends indicate that one-size-fits-all approaches are unlikely to be effective across diverse landscapes. Instead, policymakers and conservationists must consider local ecological, social, and economic factors when designing interventions to protect KBAs. The study highlights several critical priorities for biodiversity conservation in India. These include addressing the forest loss crisis in the Northeast through strengthened enforcement and community-based initiatives, learning from and replicating successful conservation models observed in states like Chhattisgarh, and developing climate-adaptive strategies to protect KBAs in the face of environmental change. Furthermore, the research emphasizes the need for international cooperation in managing transboundary ecosystems, particularly in the Himalayas and Northeast India.

The analysis also underscores the value of long-term, consistent monitoring of forest cover using remote sensing technologies. Such data is crucial for identifying trends, assessing the effectiveness of conservation measures, and guiding adaptive management of protected areas. However, it is important to note that forest cover alone is an imperfect proxy for biodiversity health, and complementary indicators and ground-based monitoring remain essential for a comprehensive understanding of ecosystem dynamics within KBAs. The study calls for a more nuanced approach to forest policy that goes beyond simple forest cover metrics to consider forest quality, ecosystem services, and the needs of local communities. This holistic perspective is essential for developing sustainable conservation strategies that balance ecological preservation with human needs.

In brief, while this analysis reveals significant challenges in protecting KBAs in India, it also provides grounds for cautious optimism. By leveraging advanced monitoring technologies, adopting targeted conservation strategies, and learning from both successes and failures, India can significantly enhance the protection of its critical biodiversity hotspots. As global environmental pressures intensify, such efforts will be crucial for preserving country's natural heritage and contributing to global biodiversity conservation and climate change mitigation efforts.

References

1. V. Jain, K.S. Rautela, M.K. Goyal, *Ecological Restoration: An Overview of Science and Policy Regime* (2023), pp. 1–27
2. N. Myers, R.A. Mittermeier, C.G. Mittermeier et al., Biodiversity hotspots for conservation priorities. Nature **403**, 853–858 (2000). https://doi.org/10.1038/35002501
3. D. Morales-Hidalgo, S.N. Oswalt, E. Somanathan, Status and trends in global primary forest, protected areas, and areas designated for conservation of biodiversity from the Global Forest Resources Assessment 2015. For. Ecol. Manage. **352**, 68–77 (2015). https://doi.org/10.1016/j.foreco.2015.06.011
4. P. Patle, P.K. Singh, S. Rakkasagi et al., *Application of Water Accounting Plus Framework for the Assessment of the Water Consumption Pattern and Food Security* (2023), pp. 257–269
5. J.A. Foley, R. DeFries, G.P. Asner et al., Global consequences of land use. Science **309**, 570–574 (2005). https://doi.org/10.1126/science.1111772
6. J.E.M. Watson, T. Evans, O. Venter et al., The exceptional value of intact forest ecosystems. Nat. Ecol. Evol. **2**, 599–610 (2018). https://doi.org/10.1038/s41559-018-0490-x
7. IUCN, *A Global Standard for the Identification of Key Biodiversity Areas (Version 1.0)* (2016)
8. R. Leberger, I.M.D. Rosa, C.A. Guerra et al., Global patterns of forest loss across IUCN categories of protected areas. Biol. Conserv. **241**, 108299 (2020). https://doi.org/10.1016/j.biocon.2019.108299
9. P.F. Donald, G.M. Buchanan, A. Balmford et al., The prevalence, characteristics and effectiveness of Aichi Target 11's "other effective area-based conservation measures" (OECMs) in Key Biodiversity Areas. Conserv. Lett. **12** (2019). https://doi.org/10.1111/conl.12659
10. R.A. Mittermeier, W.R. Turner, F.W. Larsen et al., Global biodiversity conservation: the critical role of hotspots, in *Biodiversity Hotspots*. (Springer Berlin Heidelberg, Berlin, Heidelberg, 2011), pp.3–22
11. M. Maron, J.S. Simmonds, J.E.M. Watson, Bold nature retention targets are essential for the global environment agenda. Nat. Ecol. Evol. **2**, 1194–1195 (2018). https://doi.org/10.1038/s41559-018-0595-2

12. M.K. Goyal, A.K. Gupta, S. Jha et al., Climate change impact on precipitation extremes over Indian cities: non-stationary analysis. Technol. Forecast. Soc. Change **180**, 121685 (2022). https://doi.org/10.1016/j.techfore.2022.121685
13. M.K. Goyal, A.K. Gupta, J. Das et al., Heatwave magnitude impact over Indian cities: CMIP 6 projections. Theor. Appl. Climatol. **154**, 959–971 (2023). https://doi.org/10.1007/s00704-023-04599-7
14. M.C. Hansen, P.V. Potapov, R. Moore et al., High-resolution global maps of 21st-century forest cover change. Science **342**, 850–853 (2013). https://doi.org/10.1126/science.1244693
15. P.G. Curtis, C.M. Slay, N.L. Harris et al., Classifying drivers of global forest loss. Science **361**, 1108–1111 (2018). https://doi.org/10.1126/science.aau3445
16. Global Forest Watch, *Forest Monitoring Designed for Action* (World Resources Institute, 2021). https://www.globalforestwatch.org/?ap3c=IGY4rt0QxkGqVsYGAGY4rt0DBjqj-_LwQcgGor2m-LPz1mEOXw. Accessed 31 July 2024
17. N.M. Haddad, L.A. Brudvig, J. Clobert et al., Habitat fragmentation and its lasting impact on earth's ecosystems. Sci. Adv. **1** (2015). https://doi.org/10.1126/sciadv.1500052
18. J. Barlow, F. França, T.A. Gardner et al., The future of hyperdiverse tropical ecosystems. Nature **559**, 517–526 (2018). https://doi.org/10.1038/s41586-018-0301-1
19. V. Poonia, M. Kumar Goyal, S. Jha, S. Dubey, Terrestrial ecosystem response to flash droughts over India. J. Hydrol. **605**, 127402 (2022). https://doi.org/10.1016/j.jhydrol.2021.127402
20. R. Kumar, M.K. Goyal, R.Y. Surampalli, T.C. Zhang, River pollution in India: exploring regulatory and remedial paths. Clean Technol. Environ. Policy (2024). https://doi.org/10.1007/s10098-024-02763-9
21. G.B. Bonan, Forests and climate change: forcings, feedbacks, and the climate benefits of forests. Science **320**, 1444–1449 (2008). https://doi.org/10.1126/science.1155121
22. L.V. Gatti, L.S. Basso, J.B. Miller et al., Amazonia as a carbon source linked to deforestation and climate change. Nature **595**, 388–393 (2021). https://doi.org/10.1038/s41586-021-03629-6
23. S.R. Subramoniam, S. Ravindranath, S. Rakkasagi, H. Ram, *Water Resource Management Studies at Micro Level Using Geospatial Technologies* (2022), pp. 49–74
24. J.D. Kirkpatrick, J.H. Edwards, A. Verdecchia et al., Subduction megathrust heterogeneity characterized from 3D seismic data. Nat. Geosci. **13**, 369–374 (2020). https://doi.org/10.1038/s41561-020-0562-9
25. V. Gupta, S. Rakkasagi, S. Rajpoot et al., Spatiotemporal analysis of Imja Lake to estimate the downstream flood hazard using the SHIVEK approach. Acta Geophys. (2023). https://doi.org/10.1007/s11600-023-01124-2
26. M.K. Goyal, S. Rakkasagi, S. Shaga et al., Spatiotemporal-based automated inundation mapping of Ramsar wetlands using Google Earth Engine. Sci. Rep. **13**, 17324 (2023). https://doi.org/10.1038/s41598-023-43910-4
27. S. Bhardwaj, P. Machiwar, C. Kant et al., *Analysis of Urbanization and Assessment of Its Impact on Groundwater and Land Use/Land Cover Using GIS Techniques: A Case Study of Bhopal and Gurugram District* (2023), pp. 219–255
28. J. Sayer, T. Sunderland, J. Ghazoul et al., Ten principles for a landscape approach to reconciling agriculture, conservation, and other competing land uses. Proc. Natl. Acad. Sci. **110**, 8349–8356 (2013). https://doi.org/10.1073/pnas.1210595110
29. IUCN, *Deforestation and Forest Degradation* (IUCN, 2021)
30. J. Alroy, Effects of habitat disturbance on tropical forest biodiversity. Proc. Natl. Acad. Sci. **114**, 6056–6061 (2017). https://doi.org/10.1073/pnas.1611855114
31. S.G. Potts, V. Imperatriz-Fonseca, H.T. Ngo et al., Safeguarding pollinators and their values to human well-being. Nature **540**, 220–229 (2016). https://doi.org/10.1038/nature20588
32. J. Terborgh, L. Lopez, P. Nuñez et al., Ecological meltdown in predator-free forest fragments. Science **294**, 1923–1926 (2001). https://doi.org/10.1126/science.1064397
33. S. Baidya, P. Chakraborty, S. Rakkasagi et al., Pathways to build resilience toward the impact of climate change on the Indian Sunderban, in *Ecosystem Restoration: Towards Sustainability and Resilient Development*. ed. by A.K. Gupta, M.K. Goyal, S.P. Singh (Springer Nature Singapore, Singapore, 2023), pp.307–333

References

34. S. Rakkasagi, M.K. Goyal, S. Jha, Evaluating the future risk of coastal Ramsar wetlands in India to extreme rainfalls using fuzzy logic. J. Hydrol. **632**, 130869 (2024). https://doi.org/10.1016/j.jhydrol.2024.130869
35. S. Rakkasagi, M.K. Goyal, Assessing risk levels of the extreme rainfalls in Ramsar wetlands of India using fuzzy logic, in *Chapman Conference on Remote Sensing of the Water Cycle* (AGU, 2024)
36. S. Rakkasagi, V. Poonia, M.K. Goyal, Flash drought as a new climate threat: drought indices, insights from a study in India and implications for future research. J. Water Clim. Change (2023). https://doi.org/10.2166/wcc.2023.347
37. J.E. Duffy, Why biodiversity is important to the functioning of real-world ecosystems. Front. Ecol. Environ. **7**, 437–444 (2009). https://doi.org/10.1890/070195
38. M. Kuussaari, R. Bommarco, R.K. Heikkinen et al., Extinction debt: a challenge for biodiversity conservation. Trends Ecol. Evol. **24**, 564–571 (2009). https://doi.org/10.1016/j.tree.2009.04.011
39. I.A. Hatton, K.S. McCann, J.M. Fryxell et al., The predator-prey power law: biomass scaling across terrestrial and aquatic biomes. Science **349** (2015). https://doi.org/10.1126/science.aac6284
40. A. Finger, C.J. Kettle, C.N. Kaiser-Bunbury et al., Back from the brink: potential for genetic rescue in a critically endangered tree. Mol. Ecol. **20**, 3773–3784 (2011). https://doi.org/10.1111/j.1365-294X.2011.05228.x
41. A. Sharma, M.K. Goyal, Assessment of drought trend and variability in India using wavelet transform. Hydrol. Sci. J. **65**, 1539–1554 (2020). https://doi.org/10.1080/02626667.2020.1754422
42. N. Kumar, M.K. Goyal, Projected changes in monsoonal compound dry-hot extremes in India. Atmos. Res. **310**, 107605 (2024). https://doi.org/10.1016/j.atmosres.2024.107605
43. K.S. Rautela, S. Singh, M.K. Goyal, Aerosol atmospheric rivers: patterns, impacts, and societal insights. Environ. Sci. Pollut. Res. (2024). https://doi.org/10.1007/s11356-024-34625-8
44. K.S. Rautela, S. Singh, M.K. Goyal, Resilience to air pollution: a novel approach for detecting and predicting aerosol atmospheric rivers within earth system boundaries. Earth Syst. Environ. (2024). https://doi.org/10.1007/s41748-024-00421-0
45. M. Kumar Goyal, V. Poonia, V. Jain, Three decadal urban drought variability risk assessment for Indian smart cities. J. Hydrol. **625**, 130056 (2023). https://doi.org/10.1016/j.jhydrol.2023.130056
46. M.K. Goyal, A. Sharma, R.Y. Surampalli, Remote sensing and GIS applications in sustainability, in *Sustainability*, ed. by R. Surampalli, T. Zhang, M.K. Goyal et al. (2020), pp. 605–626
47. Goyal, C.S.P. Ojha, D.H. Burn, Machine learning algorithms and their application in water resources management, in *Sustainable Water Resources Management*, pp. 165–178
48. V. Poonia, M.K. Goyal, B.B. Gupta et al., Drought occurrence in different river basins of India and blockchain technology based framework for disaster management. J. Clean. Prod. **312**, 127737 (2021). https://doi.org/10.1016/j.jclepro.2021.127737
49. J. Zhao, Y. Cai, S. Wang et al., Small and oriented wheat spike detection at the filling and maturity stages based on WheatNet. Plant Phenomics **5** (2023). https://doi.org/10.34133/plantphenomics.0109
50. G.-G. Hognogi, M. Meltzer, F. Alexandrescu, L. Ștefănescu, The role of citizen science mobile apps in facilitating a contemporary digital agora. Humanit. Soc. Sci. Commun. **10**, 863 (2023). https://doi.org/10.1057/s41599-023-02358-7
51. T. Liu, Y. Zhu, Social welfare maximization in participatory smartphone sensing. Comput. Netw. **73**, 195–209 (2014). https://doi.org/10.1016/j.comnet.2014.08.014

Chapter 3
Key Biodiversity Areas and Forest Fire

3.1 Introduction

Forest fires, whether naturally occurring or human-induced, have been integral to Earth's ecosystems for many years [1]. These dynamic and frequently devastating events are imperative in shaping landscapes, affecting ecological processes, and disturbing biodiversity across different temporal and spatial scales [2, 3]. As climate change intensifies and human activities continue to change fire regimes globally [4, 5], understanding the complex relationship between forest fires and biodiversity has become increasingly important for conservation efforts, ecosystem management, and climate change mitigation strategies [6]. This chapter focuses on the impacts of forest fires on Key Biodiversity Areas (KBAs) in India, a country rich in biodiversity facing significant challenges from changing fire regimes. KBAs are sites contributing significantly to the global persistence of biodiversity, representing the most important places for nature and human wellbeing, including areas of high species richness, threatened species, and unique ecosystems [7]. KBAs cover many ecosystems in India, from the Western Ghats' tropical forests to the Himalayan mountain meadows, each with rare fire ecology and conservation challenges [8]. The impact of forest fires on KBAs is context-dependent, varying prominently based on factors such as fire intensity, frequency, seasonality, and the ecological characteristics of impacted ecosystems [9, 10]. Fires can lead to abrupt loss of life and habitat destruction within KBAs, creating chances for environmental restoration and promoting long-term biodiversity by creating diverse habitat varieties and stimulating adaptive traits in flora and fauna [11–13]. However, the increasing frequency and intensity of fires due to climate change and human activities induce many KBAs beyond their natural resilience, possibly leading to long-term biodiversity loss and ecosystem status changes [14, 15].

In India, many ecosystems within KBAs have developed with fire, progressing adaptations, and dependencies on periodic burning. For instance, certain plant species in the dry deciduous forests of central India require the heat or smoke from fires to trigger seed germination. At the same time, some animals have developed interactive or physiological adaptations to survive and even thrive in fire-prone environments [16, 17]. However, the change in natural fire regimes, climate change, and land-use modifications pose significant threats to the biodiversity and ecological integrity of KBAs in India [18–20]. The immediate impacts of forest fires on KBAs are often intense and easily detectable. High-intensity fires can cause direct mortality of plants and animals, destroy critical habitats, and disrupt ecosystem functions [21, 22]. In India, large and rapid fires may devastate even fire-adapted species within KBAs, leading to population declines and local extinctions of rare and endangered species [23, 24]. The loss of keystone species or habitat-forming organisms, such as old-growth trees in the Western Ghats, can have cascading effects throughout the ecosystem, altering species interactions and community composition [25, 26]. However, the long-term impacts of forest fires on KBAs are more complicated and can be both positive and negative. Post-fire conditions frequently experience rapid colonization by early successional species, leading to temporary increases in local biodiversity [27, 28]. The variety of burned and unburned areas created by fires can increase landscape heterogeneity within KBAs, providing diverse habitats that support different species groups [9]. Fire-induced changes in vegetation structure and composition can also create new ecological niches, potentially promoting speciation and adaptive radiation over evolutionary timescales [29].

Conservation strategies must, therefore, balance the need to maintain natural fire regimes with the imperative to protect vulnerable species and ecosystems from excessive fire damage [30, 31]. The effective management and conservation of KBAs in India amidst changing fire regimes necessitates a comprehensive understanding of fire patterns, frequencies, and impacts across various ecosystems [32]. This chapter addresses this need by analyzing forest fire events and their effects on Indian KBAs. We explore the datasets and methodologies employed for fire frequency and trend analysis, offering insights into fires' spatial and temporal patterns within these critical biodiversity areas. Our findings, detailed in the results and discussion section, highlight the most fire-prone KBAs in India, the frequency and intensity of fire events, and their potential implications for biodiversity conservation. We examine how KBAs respond to fire, considering vegetation type, climate, and human influences. Recognizing the crucial role of policy and technology in addressing the challenges posed by forest fires in KBAs, we also explore the technology–policy interface for fire prevention and management. The interface includes a discussion of innovative approaches to fire detection, monitoring, and inhibition, as well as policy measures that can help mitigate fire threats while preserving the ecological integrity of KBAs.

3.2 Impacts of Forest Fires on KBAs

Forest fires in KBAs can have instant and severe impacts on biodiversity, causing direct mortality of flora and fauna by high-intensified fires, particularly for less movable species or those with limited fire-adaptation strategies [24, 33]. In extreme cases, fires can lead to local extinctions of rare or endemic species, often a primary reason for a site designation. For example, the 2019–2020 Australian bushfires impacted 832 native vertebrate species, including 70 threatened species, with some populations pushed to the brink of extinction [34]. Fire events can significantly alter plant groups in KBAs as many fire-adapted ecosystems rely on periodic burning for regeneration and maintaining species diversity. More frequent or intense fires can support certain plant functional types over others, leading to moves in community composition and ecosystem structure [35]. In some cases, this changes fire-sensitive forest ecosystems into more fire-prone shrublands or grasslands, a process known as biome switching [36]. Forest fires can worsen habitat fragmentation within KBAs, mainly when they coincide with other forms of land-use change. Large-scale fires can obstruct species movement and gene flow, potentially isolating populations and reducing genetic diversity [37]. The variety of burned and unburned patches created by fires can also change landscape heterogeneity within KBAs. While this patchiness can sometimes enhance biodiversity by creating a variety of habitats, it can also disrupt critical ecological processes and species interactions [38].

Forest fires can have extreme consequences on ecosystem functions within KBAs, which can impact the delivery of crucial ecosystem services. One of the most significant impacts is on carbon storage and sequestration. KBAs often contain large stocks of carbon in biomass and soils, and high-intensity fires can release substantial amounts of this carbon into the atmosphere, contributing to climate change [13, 39, 40]. Hydrological processes in KBAs can also be severely affected by forest fires. The loss of vegetation cover can lead to increased surface runoff, soil erosion, and sedimentation in water bodies, leading to cascading effects on aquatic ecosystems within KBAs, changing water quality, temperature regimes, and habitat structure [41–45]. Soil ecosystems are necessary for nutrient cycling, and supporting plant communities in KBAs can be distorted by forest fires. High-intensity fires can sterilize upper soil layers, decreasing microbial biomass and modifying soil community composition [46]. While some soil organisms have adapted to fire, changes in fire regimes can exceed their resilience thresholds. Recent research has indicated that wildfires can significantly alter soil fungal communities and nitrogen cycling, potentially impacting ecosystem recovery and function [47].

The impacts of forest fires on KBAs extend beyond the abrupt burn area and can trigger a series of indirect and cascading effects. For instance, the loss of keystone species or habitat-forming organisms due to fire can have far-reaching impacts on ecosystem structure and function. In forest ecosystems, the loss of large, old-growth trees can change canopy structure, light regimes, and microclimate conditions, influencing a wide range of dependent species [25]. Post-fire landscapes in KBAs can be more vulnerable to invasion by non-native species, which can take advantage of disturbed conditions and altered competitive dynamics. These invasive species can change fuel loads and fire regimes, creating a positive feedback loop that promotes more frequent or intense fires [48], posing a significant threat to KBAs. The impacts of forest fires on KBAs can also expand to contiguous areas through various ecological connections. For instance, the loss of habitat in a KBA might force mobile species to move into adjacent areas, altering species interactions and ecosystem dynamics in these adjacent habitats. Likewise, changes in hydrological processes or nutrient cycles within a burned KBA can have downstream effects on connected ecosystems.

One significant aspect of controlling fire impacts in KBAs is the need to balance the ecological role of fire with the imperative to protect vulnerable species and ecosystems. In many fire-adapted ecosystems, complete fire inhibition can lead to fuel accumulation and subsequent high-intensity fires that exceed the adaptive capacity of indigenous species. Therefore, management strategies must often incorporate managed burning or other fuel reduction techniques to simulate natural fire regimes while reducing the risk of terrible wildfires [49]. The detection and designation of KBAs may also need to be reviewed, considering changing fire regimes and their impacts on biodiversity. Sites formerly designated as KBAs may experience significant changes due to altered fire patterns, requiring a more dynamic approach to KBA designation and management, considering potential future scenarios under climate change and changing disturbance regimes.

3.3 Datasets and Methodology for Forest Fire Analysis

3.3.1 Data Sources and Preprocessing

This study utilized the Fire Information for Resource Management System (FIRMS) dataset to analyze forest fire trends in KBAs across India [50]. The FIRMS data provides active fire locations detected by the Moderate Resolution Imaging Spectroradiometer (MODIS) sensor aboard NASA's Terra and Aqua satellites. The FIRMS dataset has a spatial resolution of 1 km and a temporal resolution of 1 day, considering the period from 2001 to 2020. Each data point represents the location of an active fire detected when the MODIS satellite passed overhead. For fires spanning multiple 1 km pixels, a new fire event is recorded for each pixel. Similarly, for fires continuing over numerous days, a new fire event is recorded for each day the fire continues. To preprocess the data, we reclassified each daily FIRMS image into a binary raster representing fire presence (1) or absence (0) for each pixel. We then produced annual summary maps showing the total number of fire events detected in each pixel over each year from 2001 to 2020. The spatial data on KBAs in India was obtained from the World Database of Key Biodiversity Areas maintained by BirdLife International on behalf of the KBA Partnership. This dataset contains polygons representing the boundaries of identified KBAs across India.

3.3.2 Google Earth Engine Implementation

We leveraged Google Earth Engine (GEE), a cloud-based geospatial processing platform, to efficiently analyze the large volume of satellite imagery across the 20-year

3.3 Datasets and Methodology for Forest Fire Analysis 39

study period. The KBA shapefile was imported into Google Earth Engine as a "FeatureCollection," with each KBA polygon assigned a unique identifier to facilitate individual site analysis. The workflow implemented in GEE consisted of the following key steps: (i) Import the FIRMS dataset and KBA boundaries into GEE, (ii) reclassify daily FIRMS data into binary fire/no-fire rasters, (iii) create annual fire count summaries, (iv) intersect fire data with KBA boundaries, and (v) calculate standardized fire counts per KBA per year. For each KBA and each year, we calculated the total number of fire events detected within the KBA boundaries. To standardize across KBAs of different sizes, we divided the total fire count by the area of each KBA to obtain the number of fire events per square kilometer per year.

3.3.3 Statistical Analysis

To assess trends in forest fire occurrence within KBAs over the 20-year study period, we conducted a statistical time series analysis on the annual standardized fire count data for each KBA. We employed the Mann–Kendall test to determine whether each KBA showed a statistically significant monotonic trend in fire occurrence over time [51]. The Mann–Kendall test is a non-parametric test that can be applied to data with any distribution, making it suitable for our fire count data, which may not follow a normal distribution. For each KBA, we tested the null hypothesis of no trend against the alternative hypotheses of (i) monotonic downward trend, (ii) monotonic upward trend, and (iii) monotonic trend in either direction. We considered trends to be statistically significant at a p-value threshold of 0.05. Based on the test results, we classified the trend for each KBA into one of four categories: (1) Significant positive trend ($p < 0.05$, upward trend), (2) significant negative trend ($p < 0.05$, downward trend), (3) non-significant positive trend ($p \geq 0.05$, upward trend), and (4) non-significant negative trend ($p \geq 0.05$, downward trend). In addition to the trend analysis, we calculated summary statistics for each KBA, including the mean annual number of fires per square kilometer.

3.3.4 Spatial and Temporal Analysis

To observe spatial patterns in forest fire trends across India, we mapped the trend classification results for all KBAs, allowing us to visually identify any regional clustering of increasing or decreasing fire trends. We also aggregated results by state to assess broad geographic patterns. We calculated the percentage of KBAs exhibiting each trend classification for each state of India. These emphasized which states had the highest proportions of KBAs with significant increasing or decreasing fire trends. We analyzed monthly fire count data to investigate potential shifts in fire seasonality over time. For each KBA, we calculated the average number of fires occurring each month over the first five years (2001–2005) and the last five years (2016–2020) of the study period. Comparing these two time periods allowed us to identify any changes in the timing of peak fire seasons within KBAs.

3.3.5 Validation and Uncertainty Assessment

We compared our results to available ground-based fire records for a subset of KBAs to validate our remote sensing-based fire analysis. We obtained fire events reports from the Forest Survey of India (FSI) report for a sample of KBAs spread across different states of India [52, 53]. For a few of these KBAs, we calculated the correlation between our satellite-derived annual fire counts and the number of fires recorded in official events reports. To assess the classification accuracy of the FIRMS active fire product, we utilized higher-resolution satellite imagery for a sample of fire events. We manually interpreted a randomly selected subset of FIRMS fire detections using 30 m Landsat imagery acquired within 1–2 days of the reported fire date. Each point was classified as an actual fire or false detection based on visible burn scars or smoke plumes in the Landsat imagery.

3.3.6 Limitations and Considerations

Several significant limitations and considerations should be noted when interpreting the results of this analysis. MODIS's 1 km spatial resolution may ignore smaller fires, potentially underestimating fire activity in areas with frequent minor burns. Cloud cover and smoke can hide fire detection, leading to underestimation in specific regions or seasons. The MODIS fire detection has a minimum size threshold, which may vary with fire intensity and viewing angle. Changes in vegetation structure or fuel moisture over time could affect detection possibility, potentially defeating trend analysis. The analysis considers each daily 1 km fire pixel as a separate event, which may overestimate the number of independent fire occurrences. The binary description of the fire/no-fire classification does not capture fire intensity or size information. While the analysis identifies trends in fire occurrence, it does not directly address the underlying causes of these trends. The study is limited to the 2001–2020 period due to MODIS data availability, which may not fully capture longer-term fire regimes or cycles operating on multidecadal timescales.

3.4 Results and Discussion of Forest Fire in Indian KBAs

3.4.1 Overall Trend Analysis

Our analysis of forest fire trends in KBAs across India shows a complex depiction that involves both increases and encouraging signs of improved fire management. Using Google Earth Engine to analyze the FIRMS dataset from 2001 to 2020, we identified several critical patterns in fire occurrence across KBAs in India (Fig. 3.1). The results show that 6.70% of KBAs showed a statistically significant positive trend in forest fires, indicating an alarming increase in fire frequency in these areas. An additional 41.62% of KBAs showed a positive trend in forest fires that were not statistically significant, suggesting that fire occurrence may also increase in these areas, though the trend is not definitive.

Conversely, 31.25% of KBAs exhibited a statistically significant negative trend in forest fires, pointing to improvements in fire management and prevention in these areas of high biodiversity importance. A further 3.18% of KBAs showed a negative trend in forest fires that was not statistically significant. The remaining 17.25% of KBAs showed no clear trend in fire occurrence over the study period. These overall trends present a combined view of fire management in KBAs in India. While it is encouraging that nearly 35% of KBAs show negative trends in forest fires (both significant and non-significant), the fact that over 48% show positive trends is cause for concern. Specifically worrying is that 6.70% of these critical biodiversity areas are experiencing significantly increasing fire occurrence rates, emphasizing areas needing urgent fire management intervention.

3.4.2 Regional Patterns

Our analysis showed notable regional patterns in forest fire trends across KBAs in India. The Northeastern states, including Arunachal Pradesh, Assam, Manipur, and Meghalaya, showed some of the most concerning trends in forest fires (Fig. 3.2). In Arunachal Pradesh, 59.47% of KBAs showed positive trends in fire occurrence, with three sites showing statistically significant increases. Manipur showed severe fire trends, with forest fires increasing in most KBAs, though many were not statistically significant. Meghalaya also showed worrying patterns, with most KBAs experiencing positive fire occurrence trends. Assam presented a mixed picture, with over 50% of sites showing a negative trend in forest fires but 8.45% exhibiting significant positive trends. These patterns in the Northeast are particularly troubling given the region's exceptional biodiversity and high concentration of endemic species. The Eastern Himalayas biodiversity hotspot, encompassing much of Northeast India, faces severe threats from increasing fire frequency. Our results suggest that even designated regional KBAs are not immune to these pressures. Several factors may be driving these high fire occurrence rates in Northeastern KBAs, including traditional

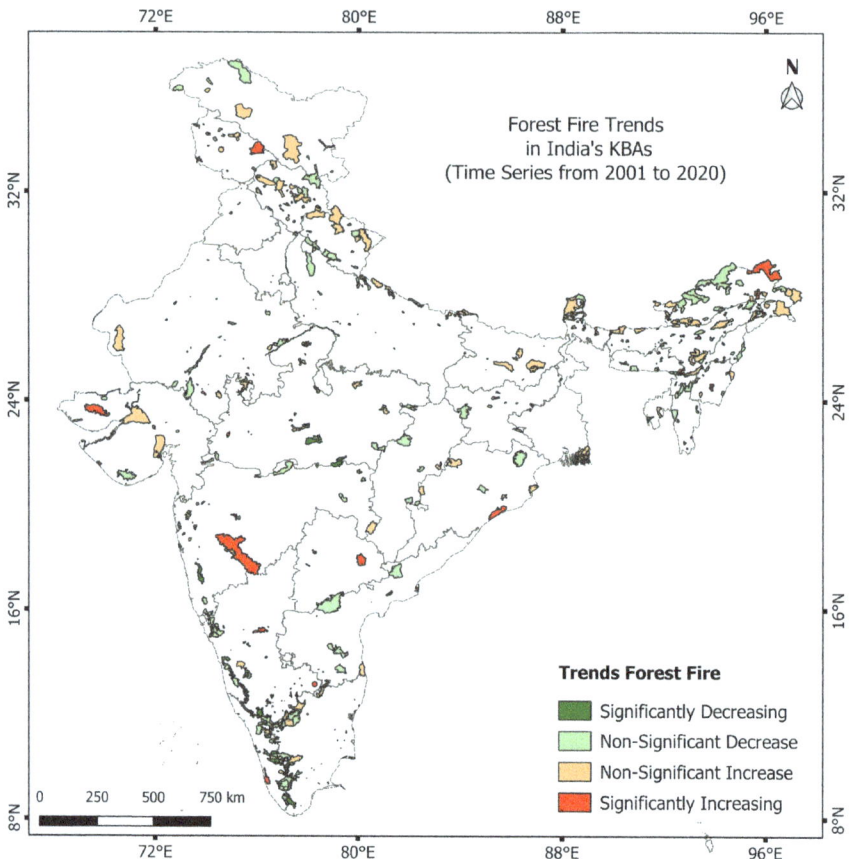

Fig. 3.1 Map visually represents the spatial distribution of forest fire trends across KBAs in India over the two-decade period (from 2001 to 2020), highlighting areas of concern and improvement in forest conservation efforts

shifting cultivation practices, extended dry seasons possibly linked to climate change, increased human access to forested areas, and inadequate fire management resources in remote areas.

The Himalayan states generally showed concerning trends (Fig. 3.3). In Himachal Pradesh, most KBAs showed an increase in forest fires, with 2.78% showing a statistically significant positive trend. Uttarakhand similarly showed mostly positive trends in fire occurrence across its KBAs. Jammu & Kashmir's KBAs showed mixed trends, with increased and decreased fire occurrence observed. These results suggest that Himalayan KBAs are facing increasing fire risks, which is particularly concerning given the sensitivity of mountain ecosystems to disturbance. Factors contributing to these trends may include climate change-induced alterations in temperature and precipitation patterns, increased human activities and tourism in some areas, and accumulation of fuel loads due to past fire inhibition policies [54–56].

3.4 Results and Discussion of Forest Fire in Indian KBAs

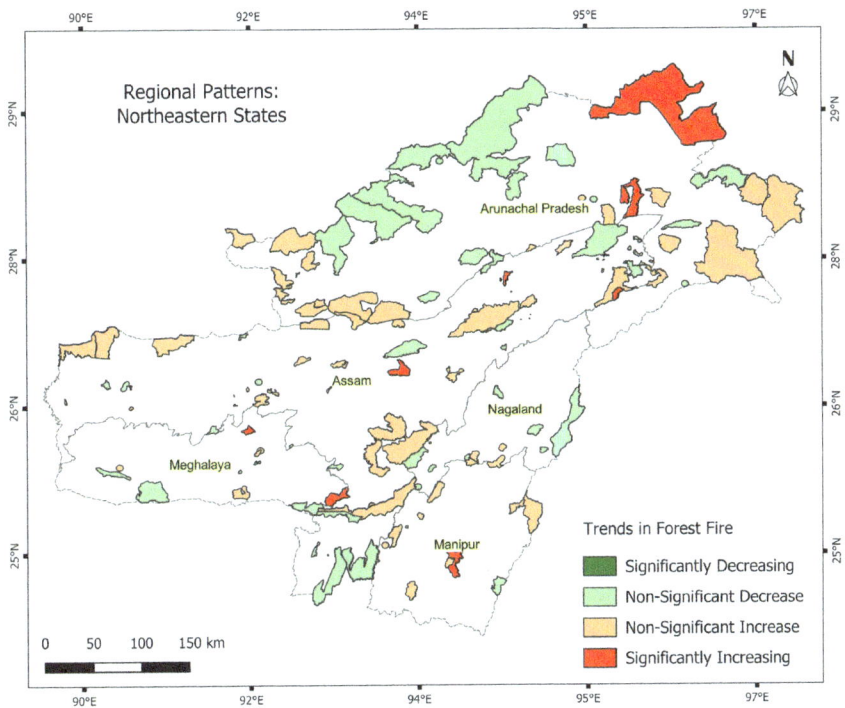

Fig. 3.2 Map visually represents the spatial distribution of forest fire trends across KBAs in Northeastern India over the two-decade period (from 2001 to 2020)

In contrast to the Northeast, several states in Western and Central India showed more positive trends. In Gujarat, forest fires did not increase significantly in any KBAs, with the majority exhibiting stable or declining fire occurrence rates. Madhya Pradesh's KBAs uniformly showed negative trends in forest fires, suggesting successful fire management interventions in the state's protected areas. Maharashtra similarly showed predominantly negative trends in fire occurrence across its KBAs except Jawaharlal Nehru Bustard Sanctuary. All of Chhattisgarh's KBAs showed negative trends in forest fires, with 77.78% exhibiting statistically significant declines, indicating a notable fire management success story. These more positive trends in central Indian states may reflect more vigorous implementation of fire prevention and management strategies, successful community-based fire management initiatives, different forest types that may be less fire-prone, and lower incidence of fire-dependent agricultural practices.

The Southern states showed mixed results. Karnataka's KBAs primarily showed negative trends in forest fires, aligning with the state's reputation for progressive forest management. Kerala exhibited a notable decline in forest fires across most sites, likely due to the state's proactive forest management policies and constructive climatic conditions. Tamil Nadu showed varied trends in fire occurrence across its

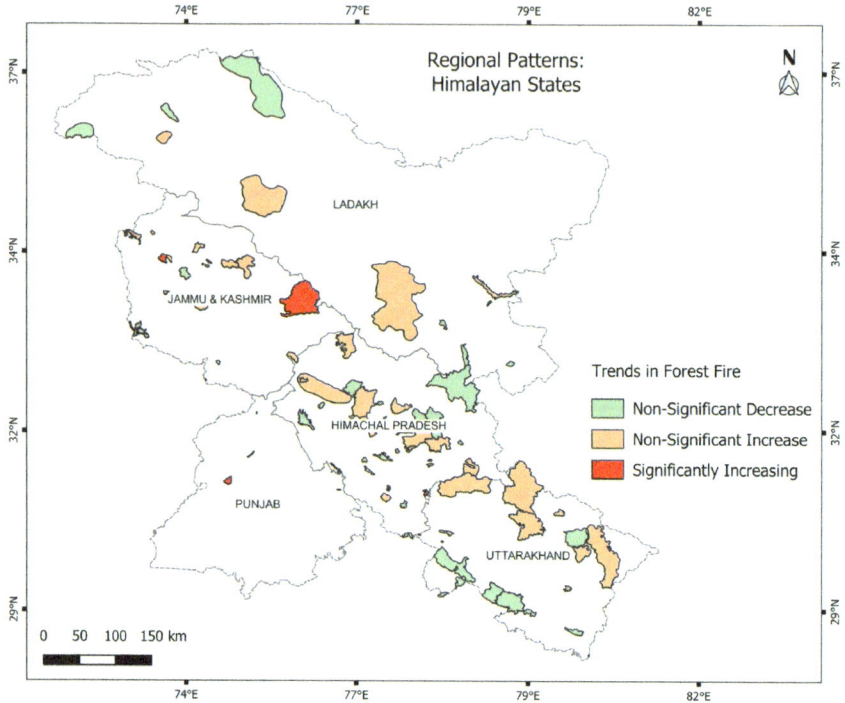

Fig. 3.3 Map visually represents the spatial distribution of forest fire trends across KBAs in the Himalayan states of India over the two-decade period (from 2001 to 2020)

KBAs, precluding any general conclusions for the state. These diverse trends in the South likely reflect the heterogeneous landscapes, varied climatic conditions, and differing anthropogenic pressures across KBAs.

3.4.3 Implications and Discussion

The heterogeneous trends in forest fire occurrence across KBAs in India have several important implications for biodiversity conservation in the country:

(a) *Targeted Interventions Needed*: The absolute regional differences in fire trends emphasize the need for targeted, context-specific fire management interventions. Given the diverse drivers of fire across India's varied landscapes, one-size-fits-all approaches are unlikely to be effective.

(b) *Northeast India Crisis*: The alarming fire incidence rates in Northeastern KBAs demand urgent attention. Strengthened fire prevention and suppression capabilities, community-based fire management initiatives, and sustainable alternatives

3.4 Results and Discussion of Forest Fire in Indian KBAs

to fire-dependent agricultural practices may all be needed to address the complex drivers of fire in the region.

(c) *Learning from Successes*: States like Kerala and Chhattisgarh demonstrate that reducing fire occurrence rates in KBAs is possible. Further research should investigate the policy measures, management practices, and local contexts enabling these fire management successes.

(d) *Climate Change Considerations*: The increasing fire trends observed in many Himalayan KBAs may be relatively responsible for climate change impacts. As temperature and precipitation patterns shift, fire management strategies in KBAs must adapt to changing fire regimes.

(e) *Ecological Implications*: Changing fire regimes can profoundly impact biodiversity and ecosystem function. In areas experiencing significant increases in fire frequency, there is a risk of ecosystem alterations, loss of fire-sensitive species, and potential degradation of habitat quality.

(f) *Importance of Long-Term Monitoring*: Our analysis demonstrates the value of consistent, long-term monitoring of forest fires using remote sensing. Such data is crucial for identifying trends, assessing management effectiveness, and guiding adaptive fire management strategies in protected areas.

(g) *Integration with Other Threat Indicators*: While our study focused on fire trends, it is essential to consider these results in the context of different threats to KBAs. For instance, our analysis found a positive correlation between increasing rates of forest fires and forest loss, suggesting compounded threats in some areas.

(h) *Policy Implications*: Our results suggest that India's fire management policies have had mixed success in protecting KBAs. While some regions show clear improvements, the increasing fire occurrence in many critical biodiversity areas suggests that more robust fire prevention and management measures may be needed.

(i) *Transboundary Considerations*: Many KBAs in India, particularly in the Northeast and Himalayas, are part of larger transboundary ecosystems. Effective long-term fire management will require international cooperation in monitoring and managing fire risks in these shared landscapes.

3.4.4 Limitations and Future Research Directions

While our analysis provides valuable insights into forest fire trends in KBAs in India, several limitations should be noted. The 1 km resolution of the FIRMS data, while suitable for national-scale analysis, may miss fine-scale fire patterns that could be ecologically significant in some KBAs. Our study focuses on fire occurrence but does not account for fire intensity, duration, or extent, as these factors influence ecological impacts. The FIRMS dataset does not differentiate between natural and human-caused fires. A more detailed analysis would be needed to attribute observed fire trends to specific drivers. Our analysis covers 2001–2020, but fire patterns may have shifted in very recent years due to policy changes or emerging pressures.

Future research directions that could build on this work include integrating multiple remote sensing products to provide a more comprehensive picture of fire impacts on biodiversity in KBAs, conducting detailed studies of select KBAs to understand local drivers of fire occurrence better and identify site-specific management needs, investigating the effectiveness of different fire management regimes in reducing fire occurrence within KBAs, assessing how fire patterns in KBAs compare to surrounding landscapes, exploring the relationships between fire occurrence trends and population changes of crucial species within KBAs, modeling future fire scenarios for KBAs under different climate change trajectories, investigating the links between fire trends and changes in human land use and livelihoods, and exploring how local knowledge about fire management could be integrated into KBA fire management strategies.

3.5 Technology–Policy Interface to Prevent and Manage Forest Fires

The effective prevention and management of forest fires in KBAs require an advanced integration of cutting-edge technology and well-crafted policies. This technology–policy interface is crucial for addressing the complex challenges posed by changing fire regimes, climate change, and human activities (Fig. 3.4). One of the primary technological advancements in forest fire management is the use of satellite-based remote sensing systems, such as NASA's MODIS and the VIIRS, providing near-real-time fire detection capabilities. By continuously monitoring large areas, these satellites can identify active fires, even in remote or inaccessible areas. Integrating this technology into national and regional fire management strategies requires supportive policies that ensure data accessibility, promote data sharing across agencies, and allocate resources to interpret and act on satellite data. Building on satellite data, advanced fire prediction models have been developed that integrate various environmental parameters such as weather conditions, vegetation type, and topography. These models can predict fire risk with increasing accuracy, allowing for proactive measures to be taken in high-risk areas. To fully leverage these technological capabilities, policies that mandate using such predictive tools in forest management planning and resource allocation must be in place.

Unmanned Aerial Vehicles (UAVs) or drones have emerged as a valuable tool for fire detection and monitoring [57]. The drones can survey large areas quickly, detect hot spots, and provide real-time information to firefighting teams, equipped with thermal imaging cameras. In the context of KBAs, drones can be beneficial for monitoring remote or rugged areas that are difficult to access on foot. However, the widespread adoption of drone technology for fire management requires carefully crafted policies that address issues such as airspace regulations, privacy concerns, and operator training and certification. Policies should also encourage the development of procedures for integrating drone-collected data into existing fire management

3.5 Technology–Policy Interface to Prevent and Manage Forest Fires

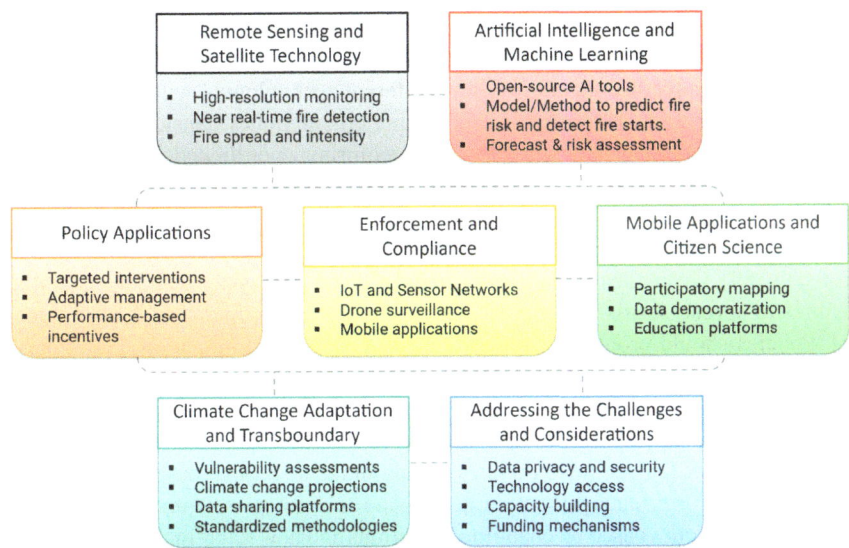

Fig. 3.4 Diagram outlines different technological tools and approaches that can be integrated to support the technology–policy interface to reduce forest fires, from data analysis to policy implementation in KBAs

systems. Artificial intelligence (AI) and machine learning (ML) are increasingly applied to various aspects of forest fire management [58]. These technologies can analyze data from multiple sources to improve fire prediction accuracy, optimize resource allocation, and even assist in post-fire damage assessment. For instance, ML algorithms can be trained on historical fire data to identify patterns and predict future fire behavior more precisely. The policies that encourage data collection and sharing must be developed, promote the development and validation of AI models, and provide guidelines for ethical use in decision-making processes. On-the-ground sensor networks represent another technological frontier in forest fire management. These networks consist of strategically placed sensors that monitor environmental conditions such as temperature, humidity, and wind speed [14]. When integrated with communication systems, these sensors can provide early warning of fire-conducive conditions or detect fires in their early stages. Implementing such sensor networks in KBAs could significantly enhance early detection capabilities. However, deploying these systems requires policies addressing environmental impact, maintenance responsibilities, and data management. Policies should also promote standardizing sensor technologies and data formats to ensure interoperability across different systems and regions.

While these technological advancements offer potent tools for forest fire management, their practical implementation requires a supportive policy environment. One critical aspect is the development of comprehensive fire management plans for each KBA that incorporate these technologies. These plans should be mandated by policy and regularly updated to reflect technological advancements and changing environmental conditions. They should outline straightforward fire prevention, detection, and response protocols, specifying how different technologies will be utilized at each stage. Inter-agency coordination is another crucial policy consideration. Forest fires often cross administrative boundaries, requiring coordinated responses from multiple agencies. Policies should establish clear frameworks for information sharing and joint operations among government departments and state and national agencies, including creating centralized data platforms and standardized communication protocols [59]. Given the transboundary nature of many ecosystems and the potential for fires to spread across national borders, international cooperation is also essential. Policies should promote cross-border fire management initiatives, particularly for KBAs that span international boundaries, involving joint monitoring systems, sharing early warning mechanisms, and coordinating response protocols.

Adaptive management becomes increasingly important as climate change continues to alter fire regimes. Policies should mandate regular assessments of the effectiveness of fire management strategies and technologies, with provisions for adjusting approaches based on changing conditions and new scientific understanding. Data management and sharing policies are crucial for maximizing the benefits of fire management technologies. Clear guidelines should be established for data collection, storage, access, and sharing, ensuring that valuable information is available to all relevant stakeholders while protecting sensitive information. Policies should also address data standardization issues to facilitate integration and analysis across different platforms and agencies.

In brief, the policies should be updated regularly to reflect the capabilities and limitations of new technologies. For example, as satellite-based fire detection systems improve accuracy and temporal resolution, response times and resource allocation policies should be adjusted accordingly. Government policies can drive the adoption of new technologies through funding initiatives, regulatory requirements, and standardization efforts. For instance, policies mandating specific fire risk assessment methodologies can encourage the development of AI-based risk prediction tools [58]. Establish platforms that bring together technology developers, policymakers, fire management professionals, and researchers to facilitate knowledge exchange and collaborative problem-solving. Implement adaptive management approaches that rapidly integrate new technologies and scientific findings into fire management strategies. Establish technical standards and protocols to ensure fire management technologies and data systems interoperability, supported by policies that mandate adherence to these standards in government procurement and funding decisions. Implement regular technology impact assessments to evaluate new fire management technologies' effectiveness and unintended consequences. Use these assessments to inform policy adjustments and future technology investments.

3.6 Conclusion

This comprehensive analysis of forest fire trends in KBAs across India from 2001 to 2020 reveals a complex and heterogeneous landscape of fire regimes and management challenges. The study's findings highlight both areas of concern and instances of successful fire management, underscoring the need for targeted, context-specific interventions to protect these critical biodiversity hotspots. The alarming increase in fire occurrence in 6.70% of KBAs and positive trends in 41.62% of sites signals a pressing need for enhanced fire prevention and management strategies in many areas. The Northeastern states and parts of the Himalayas are particularly concerned, where many KBAs exhibited increasing fire trends. These regions, known for their exceptional biodiversity and high concentration of endemic species, face compounded threats from changing fire regimes, climate change, and human activities.

Conversely, the study also revealed encouraging trends, with 31.25% of KBAs showing statistically significant decreases in fire occurrence. States like Kerala, Chhattisgarh, and parts of central India demonstrated notable success in reducing fire events, offering valuable lessons for effective fire management practices that could be adapted and implemented in other regions. The observed spatial variability in fire trends underscores the importance of tailored management approaches that account for local ecological, climatic, and socioeconomic factors. Given the diverse drivers of fire across India's varied landscapes, one-size-fits-all policies are unlikely to be effective. Instead, a nuanced understanding of regional fire dynamics and locally appropriate interventions is essential for effective biodiversity conservation in KBAs.

The study's findings have significant implications for policy and management. They highlight the need for urgent interventions in high-risk areas, particularly in the Northeast and Himalayas, and enhanced monitoring and early warning systems that leverage advanced remote sensing and AI technologies. Strengthened inter-agency coordination and transboundary cooperation are crucial for effective fire management, as is integrating climate change considerations into long-term strategies. While this analysis provides valuable insights, it also reveals areas for future research. More detailed studies on the ecological impacts of changing fire regimes, the effectiveness of different management strategies, and the interplay between fire and other threats to biodiversity are needed. Additionally, finer-scale analyses that account for fire intensity and duration could provide a more nuanced understanding of fire impacts on KBA ecosystems. Such research would further enhance our ability to develop targeted and effective conservation strategies.

References

1. V. Jain, K.S. Rautela, M.K. Goyal, *Ecological Restoration: An Overview of Science and Policy Regime* (2023), pp. 1–27

2. V. Gupta, S. Rakkasagi, S. Rajpoot et al., Spatiotemporal analysis of Imja Lake to estimate the downstream flood hazard using the SHIVEK approach. Acta Geophys. (2023). https://doi.org/10.1007/s11600-023-01124-2
3. J.G. Pausas, J.E. Keeley, Wildfires as an ecosystem service. Front. Ecol. Environ. **17**, 289–295 (2019). https://doi.org/10.1002/fee.2044
4. K.S. Rautela, S. Singh, M.K. Goyal, Aerosol atmospheric rivers: patterns, impacts, and societal insights. Environ. Sci. Pollut. Res. (2024). https://doi.org/10.1007/s11356-024-34625-8
5. K.S. Rautela, S. Singh, M.K. Goyal, Resilience to air pollution: a novel approach for detecting and predicting aerosol atmospheric rivers within earth system boundaries. Earth Syst. Environ. (2024). https://doi.org/10.1007/s41748-024-00421-0
6. D.M.J.S. Bowman, C.A. Kolden, J.T. Abatzoglou et al., Vegetation fires in the Anthropocene. Nat. Rev. Earth Environ. **1**, 500–515 (2020). https://doi.org/10.1038/s43017-020-0085-3
7. IUCN, *A Global Standard for the Identification of Key Biodiversity Areas (Version 1.0)* (2016)
8. N. Kodandapani, M.A. Cochrane, R. Sukumar, Forest fire regimes and their ecological effects in seasonally dry tropical ecosystems in the Western Ghats, India, in *Tropical Fire Ecology*. (Springer Berlin Heidelberg, Berlin, Heidelberg, 2009), pp.335–354
9. L.T. Kelly, L. Brotons, Using fire to promote biodiversity. Science **355**, 1264–1265 (2017). https://doi.org/10.1126/science.aam7672
10. S.R. Levick, G.P. Asner, I.P.J. Smit, Spatial patterns in the effects of fire on savanna vegetation three-dimensional structure. Ecol. Appl. **22**, 2110–2121 (2012). https://doi.org/10.1890/12-0178.1
11. S. Baidya, P. Chakraborty, S. Rakkasagi et al., Pathways to build resilience toward the impact of climate change on the Indian Sunderban, in *Ecosystem Restoration: Towards Sustainability and Resilient Development*. ed. by A.K. Gupta, M.K. Goyal, S.P. Singh (Springer Nature Singapore, Singapore, 2023), pp.307–333
12. T. He, B.B. Lamont, J.G. Pausas, Fire as a key driver of earth's biodiversity. Biol. Rev. **94**, 1983–2010 (2019). https://doi.org/10.1111/brv.12544
13. C.M. Gibson, L.E. Chasmer, D.K. Thompson et al., Wildfire as a major driver of recent permafrost thaw in boreal peatlands. Nat. Commun. **9**, 3041 (2018). https://doi.org/10.1038/s41467-018-05457-1
14. C.S. Stevens-Rumann, K.B. Kemp, P.E. Higuera et al., Evidence for declining forest resilience to wildfires under climate change. Ecol. Lett. **21**, 243–252 (2018). https://doi.org/10.1111/ele.12889
15. J.L. Baltzer, N.J. Day, X.J. Walker et al., Increasing fire and the decline of fire adapted black spruce in the boreal forest. Proc. Natl. Acad. Sci. **118** (2021). https://doi.org/10.1073/pnas.2024872118
16. M.S. Sritharan, B.C. Scheele, W. Blanchard et al., Plant rarity in fire-prone dry sclerophyll communities. Sci. Rep. **12**, 12055 (2022). https://doi.org/10.1038/s41598-022-15927-8
17. J.G. Pausas, Evolutionary fire ecology: lessons learned from pines. Trends Plant Sci. **20**, 318–324 (2015). https://doi.org/10.1016/j.tplants.2015.03.001
18. S. Bhardwaj, P. Machiwar, C. Kant et al., *Analysis of Urbanization and Assessment of Its Impact on Groundwater and Land Use/Land Cover Using GIS Techniques: A Case Study of Bhopal and Gurugram District* (2023), pp. 219–255
19. S. Rakkasagi, M.K. Goyal, Assessing risk levels of the extreme rainfalls in Ramsar wetlands of India using fuzzy logic, in *Chapman Conference on Remote Sensing of the Water Cycle* (AGU, 2024)
20. C. Sudhakar Reddy, C.S. Jha, G. Manaswini et al., Nationwide assessment of forest burnt area in India using Resourcesat-2 AWiFS data. Curr. Sci. **112**, 1521 (2017). https://doi.org/10.18520/cs/v112/i07/1521-1532
21. S.M. Juárez-Orozco, C. Siebe, D. Fernández y Fernández, Causes and effects of forest fires in tropical rainforests: a bibliometric approach. Trop. Conserv. Sci. **10**, 194008291773720 (2017). https://doi.org/10.1177/1940082917737207
22. J.G. Canadell, D.E. Pataki, L.F. Pitelka, *Terrestrial Ecosystems in a Changing World* (Springer Berlin Heidelberg, Berlin, Heidelberg, 2007)

References

23. D.G. Nimmo, S. Avitabile, S.C. Banks et al., Animal movements in fire-prone landscapes. Biol. Rev. **94**, 981–998 (2019). https://doi.org/10.1111/brv.12486
24. R.C. Godfree, N. Knerr, F. Encinas-Viso et al., Implications of the 2019–2020 megafires for the biogeography and conservation of Australian vegetation. Nat. Commun. **12**, 1023 (2021). https://doi.org/10.1038/s41467-021-21266-5
25. D.B. Lindenmayer, R.M. Kooyman, C. Taylor et al., Recent Australian wildfires made worse by logging and associated forest management. Nat. Ecol. Evol. **4**, 898–900 (2020). https://doi.org/10.1038/s41559-020-1195-5
26. N.J. Day, K.E. Dunfield, J.F. Johnstone et al., Wildfire severity reduces richness and alters composition of soil fungal communities in boreal forests of western Canada. Glob. Change Biol. **25**, 2310–2324 (2019). https://doi.org/10.1111/gcb.14641
27. L.A. Burkle, J.A. Myers, R.T. Belote, Wildfire disturbance and productivity as drivers of plant species diversity across spatial scales. Ecosphere **6**, 1–14 (2015). https://doi.org/10.1890/ES15-00438.1
28. A.I.T. Tulloch, J. Pichancourt, C.R. Gosper et al., Fire management strategies to maintain species population processes in a fragmented landscape of fire-interval extremes. Ecol. Appl. **26**, 2175–2189 (2016). https://doi.org/10.1002/eap.1362
29. T. He, B.B. Lamont, Baptism by fire: the pivotal role of ancient conflagrations in evolution of the earth's flora. Natl. Sci. Rev. **5**, 237–254 (2018). https://doi.org/10.1093/nsr/nwx041
30. L.T. Kelly, K.M. Giljohann, A. Duane et al., Fire and biodiversity in the Anthropocene. Science **370** (2020). https://doi.org/10.1126/science.abb0355
31. J.E. Keeley, J.G. Pausas, P.W. Rundel et al., Fire as an evolutionary pressure shaping plant traits. Trends Plant Sci. **16**, 406–411 (2011). https://doi.org/10.1016/j.tplants.2011.04.002
32. R. Kumar, M.K. Goyal, R.Y. Surampalli, T.C. Zhang, River pollution in India: exploring regulatory and remedial paths. Clean Technol. Environ. Policy (2024). https://doi.org/10.1007/s10098-024-02763-9
33. N. Kumar, M.K. Goyal, Projected changes in monsoonal compound dry-hot extremes in India. Atmos. Res. **310**, 107605 (2024). https://doi.org/10.1016/j.atmosres.2024.107605
34. B.A. Wintle, S. Legge, J.C.Z. Woinarski, After the megafires: what next for Australian wildlife? Trends Ecol. Evol. **35**, 753–757 (2020). https://doi.org/10.1016/j.tree.2020.06.009
35. R.G. Miller, R. Tangney, N.J. Enright et al., Mechanisms of fire seasonality effects on plant populations. Trends Ecol. Evol. **34**, 1104–1117 (2019). https://doi.org/10.1016/j.tree.2019.07.009
36. J.G. Pausas, W.J. Bond, Alternative biome states in terrestrial ecosystems. Trends Plant Sci. **25**, 250–263 (2020). https://doi.org/10.1016/j.tplants.2019.11.003
37. D.B. Lindenmayer, C. Sato, Hidden collapse is driven by fire and logging in a socioecological forest ecosystem. Proc. Natl. Acad. Sci. **115**, 5181–5186 (2018). https://doi.org/10.1073/pnas.1721738115
38. L.T. Kelly, A.F. Bennett, M.F. Clarke, M.A. McCarthy, Optimal fire histories for biodiversity conservation. Conserv. Biol. **29**, 473–481 (2015). https://doi.org/10.1111/cobi.12384
39. M. Kumar Goyal, V. Poonia, V. Jain, Three decadal urban drought variability risk assessment for Indian smart cities. J. Hydrol. **625**, 130056 (2023). https://doi.org/10.1016/j.jhydrol.2023.130056
40. V. Poonia, M. Kumar Goyal, S. Jha, S. Dubey, Terrestrial ecosystem response to flash droughts over India. J. Hydrol. **605**, 127402 (2022). https://doi.org/10.1016/j.jhydrol.2021.127402
41. P. Patle, P.K. Singh, S. Rakkasagi et al., *Application of Water Accounting Plus Framework for the Assessment of the Water Consumption Pattern and Food Security* (2023), pp. 257–269
42. S.R. Subramoniam, S. Ravindranath, S. Rakkasagi, H. Ram, *Water Resource Management Studies at Micro Level Using Geospatial Technologies* (2022), pp. 49–74
43. S. Rakkasagi, V. Poonia, M.K. Goyal, Flash drought as a new climate threat: drought indices, insights from a study in India and implications for future research. J. Water Clim. Change (2023). https://doi.org/10.2166/wcc.2023.347
44. M.K. Goyal, A.K. Gupta, J. Das et al., Heatwave magnitude impact over Indian cities: CMIP 6 projections. Theor. Appl. Climatol. **154**, 959–971 (2023). https://doi.org/10.1007/s00704-023-04599-7

45. F.-N. Robinne, D.W. Hallema, K.D. Bladon, J.M. Buttle, Wildfire impacts on hydrologic ecosystem services in North American high-latitude forests: a scoping review. J. Hydrol. **581**, 124360 (2020). https://doi.org/10.1016/j.jhydrol.2019.124360
46. M.R.A. Pingree, L.N. Kobziar, The myth of the biological threshold: a review of biological responses to soil heating associated with wildland fire. For. Ecol. Manage. **432**, 1022–1029 (2019). https://doi.org/10.1016/j.foreco.2018.10.032
47. A.R. Nelson, A.B. Narrowe, C.C. Rhoades et al., Wildfire-dependent changes in soil microbiome diversity and function. Nat. Microbiol. **7**, 1419–1430 (2022). https://doi.org/10.1038/s41564-022-01203-y
48. J.K. Balch, B.A. Bradley, C.M. D'Antonio, J. Gómez-Dans, Introduced annual grass increases regional fire activity across the arid western USA (1980–2009). Glob. Change Biol. **19**, 173–183 (2013). https://doi.org/10.1111/gcb.12046
49. S.L. Stephens, M.A. Battaglia, D.J. Churchill et al., Forest restoration and fuels reduction: convergent or divergent? Bioscience (2020). https://doi.org/10.1093/biosci/biaa134
50. C. Yue, P. Ciais, P. Cadule et al., Modelling the role of fires in the terrestrial carbon balance by incorporating SPITFIRE into the global vegetation model ORCHIDEE—part 1: simulating historical global burned area and fire regimes. Geosci. Model Dev. **7**, 2747–2767 (2014). https://doi.org/10.5194/gmd-7-2747-2014
51. A. Sharma, M.K. Goyal, Assessment of drought trend and variability in India using wavelet transform. Hydrol. Sci. J. **65**, 1539–1554 (2020). https://doi.org/10.1080/02626667.2020.1754422
52. Forest Survey of India, *India State of Forest Report 2011* (2011)
53. ISFR, *India State of Forest Report* (2019)
54. M.K. Goyal, A.K. Gupta, S. Jha et al., Climate change impact on precipitation extremes over Indian cities: non-stationary analysis. Technol. Forecast. Soc. Change **180**, 121685 (2022). https://doi.org/10.1016/j.techfore.2022.121685
55. M.K. Goyal, S. Rakkasagi, S. Shaga et al., Spatiotemporal-based automated inundation mapping of Ramsar wetlands using Google Earth Engine. Sci. Rep. **13**, 17324 (2023). https://doi.org/10.1038/s41598-023-43910-4
56. S. Rakkasagi, M.K. Goyal, S. Jha, Evaluating the future risk of coastal Ramsar wetlands in India to extreme rainfalls using fuzzy logic. J. Hydrol. **632**, 130869 (2024). https://doi.org/10.1016/j.jhydrol.2024.130869
57. B. Majidi, O. Hemmati, F. Baniardalan et al., Geo-spatiotemporal intelligence for smart agricultural and environmental eco-cyber-physical systems, in *Enabling AI Applications in Data Science*. ed. by A.-E. Hassanien, M.H.N. Taha, N.E.M. Khalifa (Springer International Publishing, Cham, 2021), pp.471–491
58. S. Singh, M.K. Goyal, Enhancing climate resilience in businesses: the role of artificial intelligence. J. Clean. Prod. **418**, 138228 (2023). https://doi.org/10.1016/j.jclepro.2023.138228
59. M.M. Steen-Adams, S. Charnley, M.D.O. Adams, Cross-boundary cooperation in wildfire management during the custodial management period of the US Forest Service: a case study of the eastern Cascades of Oregon, USA, 1905–1945. Land Use Policy **127**, 106550 (2023). https://doi.org/10.1016/j.landusepol.2023.106550

Chapter 4
Key Biodiversity Areas and Stable Night Lights

4.1 Introduction

Stable night light datasets have developed as a robust remote sensing tool for monitoring and assessing human activity and development across large geographic areas. As artificial lighting at night has extended intensely over the past several decades, satellite-based observations of night lights have provided exceptional insights into human settlement patterns, economic activity, and infrastructure development [1]. Within the context of biodiversity conservation, stable night light data presents valuable information about human-caused pressures and threats to critical habitats and species. Night light remote sensing depends primarily on satellite sensors that detect visible and near-infrared emissions from the Earth's surface during night-time hours. The longest-running and most widely used dataset comes from the US Air Force Defense Meteorological Satellite Program's Operational Linescan System (DMSP OLS), which has collected global night light observations since the 1970s [2]. More recently, the Visible Infrared Imaging Radiometer Suite (VIIRS) Day/Night Band has provided higher-resolution night light data since 2012 [1]. These sensors detect night-time light sources, including city lights, gas flares, fires, moonlit clouds, and terrain.

Stable night lights—persistent light emissions from cities, towns, and other fixed infrastructure—are of primary interest for ecological and conservation applications. Various processing techniques have been developed to filter out ephemeral lights and background noise to produce stable light products [3]. These stable light datasets provide a consistent, objective measure of human presence and activity that can be tracked over time and compared across regions. In the context of Key Biodiversity Areas (KBAs) and other critical habitats [4, 5], stable night light data offers several key advantages as an indicator of human-caused pressure. These include global coverage, frequent observations, long-time series, correlation with development, sensitivity to change, and cloud penetration capabilities. These characteristics make stable night lights a valuable complement to other remote sensing indicators like forest cover change or land-use classification. Night lights provide a more direct measure of human presence and activity level that can reveal pressures on natural habitats even before land cover changes are apparent. Numerous studies have demonstrated the utility of night light data for conservation applications. Aubrecht et al. used DMSP OLS night lights to assess human influence on protected areas globally, finding that light pollution affects a significant portion of protected lands [6]. Sánchez de Miguel et al. analyzed trends in night light emissions within and around protected areas, revealing increasing development pressure in many regions [7]. More recently, Beyer et al. incorporated night light data into a comprehensive assessment of human pressure on the world's protected areas [8].

For KBAs specifically, night lights offer insights into several vital threats and pressures, including urban expansion, infrastructure development, resource extraction, agricultural intensification, and tourism development [9, 10]. Recent studies have further highlighted the value of night light data in conservation contexts. For example, Huang et al. used VIIRS night light data to assess urbanization impacts on global protected areas, finding significant increases in light pollution within and around many reserves [11]. Gaston et al. reviewed the latest evidence on the ecological impacts of artificial light at night, emphasizing its pervasive effects on biodiversity and ecosystem functions [12]. Beyond indicating physical encroachment and land-use change, artificial lighting can negatively impact species and ecosystems within KBAs. A growing body of research has documented wide-ranging ecological effects of light pollution, including disruption of animal behavior and movement patterns, interference with circadian rhythms and physiological processes, disorientation of nocturnal species and migratory birds, alteration of predator–prey interactions and community composition, and changes to plant growth, phenology, and pollination [13–15]. These impacts highlight the importance of monitoring night light trends as an indirect indicator of human activity and as a direct pressure on biodiversity. As Gaston et al. argue, artificial light at night should be considered a driver of global change on par with factors like habitat loss and climate change [13]. Recent work by Seymoure has further emphasized the need to consider the spectral composition of artificial lighting, as different organisms respond differently to various wavelengths of light [16].

The present study analyzes trends in stable night lights for KBAs across India using two complementary datasets: the DMSP OLS Night-time Lights Time Series (1992–2013) and VIIRS Day/Night Band Composites (2014–2020). Combining these datasets, we aim to characterize long-term trajectories and recent dynamics of night light change in and around KBAs in India. The DMSP OLS data allows us to examine two decades of night light trends leading up to 2013, capturing the rapid development across much of India during this period. The VIIRS data then provides a higher-resolution view of recent changes from 2014 to 2020, offering insights into current pressures and emerging threats. Therefore, we analyze trends within each dataset separately rather than attempting to create a continuous time series. For each KBA, we extract annual mean night light values from both datasets and apply the Mann–Kendall trend test to assess the significance and direction of trends in night light intensity over the two time periods (1992–2013 and 2014–2020). The results of this analysis can help identify KBAs facing the most significant development pressure and light pollution impacts, informing conservation prioritization and management strategies.

4.2 Effects of Artificial Light on Wildlife

This study's long-term trajectories and recent dynamics of night light change can help us understand the impacts on various species and ecosystems within KBAs in India [17]. Here, we discuss the potential effects of increasing artificial night lights on wildlife, contextualized within the framework of this research. The increasing trends in night light intensity observed in many KBAs could significantly disrupt the circadian rhythms of various species. For instance, the Indian flying fox, an important pollinator and seed disperser, may alter its foraging patterns in response to increased night light. A study in urban areas of India found that artificial lighting affected these bats' emergence time and foraging duration [18]. The long-term data from DMSP OLS (1992–2013) could reveal gradual changes in habitat use by such species, while the more recent high-resolution VIIRS data (2014–2020) might highlight areas where these effects have intensified.

Many of KBAs in India are crucial breeding grounds for various species. The observed trends in night light intensity could have significant implications for reproductive behaviors and success rates. For example, the Olive Ridley Sea turtles, which nest on the beaches of Odisha, are susceptible to artificial light. Increased coastal illumination, potentially captured in the VIIRS dataset, could disorient nesting females and emerging hatchlings, reducing reproductive success [19]. Similarly, for bird species in KBAs, such as the Great Indian Bustard or the Bengal Florican, increasing night lights could alter mating calls and breeding behaviors, as observed in other avian species globally [20]. The spatial analysis of night light trends can reveal the potential fragmentation of habitats within and around KBAs. Brightly lit areas can act as barriers to movement for species sensitive to light, effectively fragmenting populations [21]. In the Indian context, this could be particularly problematic for species like elephants or leopards that require large, connected habitats. KBAs showing significant increases in night light intensity, especially those identified through the Mann–Kendall trend test, might indicate areas where wildlife corridors are being compromised. The long-term DMSP OLS data could show gradual fragmentation effects, while the VIIRS data might highlight more recent and acute barriers.

Chronic exposure to night light can induce physiological stress in wildlife, affecting overall health and fitness. In Indian KBAs experiencing increasing light pollution, species might suffer from altered hormone levels, reduced immune function, and other health issues [22–24]. For example, light-sensitive amphibians in the Western Ghats, a biodiversity hotspot, could experience increased stress and disease susceptibility [25]. The temporal resolution of both datasets allows for identifying KBAs where wildlife might be facing long-term, chronic exposure to night light, potentially informing health monitoring programs for critical species. Many KBAs in India are known for their rich floral diversity and complex plant–pollinator interactions. Increasing night lights, as potentially revealed by the trend analysis, could disrupt these delicate relationships. Night-blooming plants might flower at inappropriate times, and nocturnal pollinators like moths could be disoriented or deterred by artificial light, cascading effects on ecosystem functions and services. The spatial patterns of night light change across KBAs could help identify areas where such disruptions are most likely to occur, guiding conservation strategies for maintaining healthy plant–pollinator networks. KBAs in India are essential stopover sites and destinations for numerous migratory species, including birds and marine animals. The trends in night light intensity, especially in coastal KBAs, could have significant implications for these migratory patterns. Artificial light can disorient migrating birds, leading to collisions with structures or depletion of energy reserves. The VIIRS data, with its higher resolution, might be instrumental in identifying recent changes in coastal illumination that could affect these migratory routes and behaviors.

4.3 Datasets and Methodology for Monitoring Night-Time Lighting

4.3.1 Data Sources and Preprocessing

Night-time light data has proven to be a powerful indicator of global human activity and development. This study utilizes two extensive datasets to examine lighting trends within KBAs, shedding light on the complex relationship between human expansion and crucial biodiversity habitats. The study employs the Defense Meteorological Program Operational Line Scan System (DMSP OLS) Night-time Lights Time Series (1992–2013) at 30 arc-second resolution [3]. Its "stable lights" band is instrumental, excluding temporary light sources and background noise, focusing on persistent illumination typically associated with human settlements and infrastructure. Thus, this study incorporates the recent Visible Infrared Imaging Radiometer Suite (VIIRS) Stray Light Corrected Night-time Day/Night Band Composites (2014–2020), offering higher resolution at 15 arc seconds [1]. Unlike DMSP OLS, VIIRS captures all detected light, including background noise and transient events, providing a more comprehensive view of night-time illumination but presenting challenges in low-light areas. Data preprocessing involved creating annual composite images from the DMSP OLS "stable lights" band and processing weekly VIIRS composites into annual average radiance images. This meticulous preparation ensures data consistency while accounting for the data's methodological differences, laying a solid foundation for subsequent analysis.

4.3 Datasets and Methodology for Monitoring Night-Time Lighting   57

4.3.2 Google Earth Engine Implementation

While not explicitly mentioned in the original text, implementing this analysis would benefit significantly from using Google Earth Engine (GEE) [26, 27]. GEE is a cloud-based platform that allows the processing large-scale geospatial datasets, making it ideal for handling the global night-time lighting data used in this study. Researchers could efficiently process and analyze the DMSP OLS and VIIRS datasets using GEE across time series. The platform's ability to handle large volumes of satellite imagery and perform computations in the cloud would greatly expedite the creation of annual composites and the extraction of light values for each KBA. GEE's JavaScript is used for the preprocessing steps, including aggregating nightly observations into yearly summaries for DMSP OLS data and creating annual average radiance images from weekly composites for VIIRS data. Furthermore, GEE's integration with vector data would facilitate the intersection of the processed night-time lighting rasters with the KBA shapefile, allowing for rapid extraction of average light values for each KBA across all years of the study period.

4.3.3 Statistical Analysis

The trend detection is performed on the resulting time series data [28, 29]. For each KBA, two-time series were created: one for the DMSP OLS stable lights (1992–2013) and another for the VIIRS average radiance (2014–2020). This separation is crucial due to the methodological differences between the two datasets, which preclude direct comparison of absolute values. The study employed the Mann–Kendall test to detect significant trends in these time series. This non-parametric statistical test is well-suited for detecting monotonic trends in time series data, making it an appropriate choice for analyzing changes in night-time lighting over time. The test was applied to each KBA's time series, with a significance level set at $p < 0.05$. This rigorous statistical approach identifies trends unlikely to have occurred by chance, providing a solid foundation for interpreting the results. Based on the outcomes of the Mann–Kendall tests, trends for each KBA were categorized into four groups: significant positive trend, positive trend (non-significant), negative trend (non-significant), and significant negative trend. This categorization allows for a nuanced understanding of the changes occurring in different KBAs.

4.3.4 Spatial and Temporal Analysis

The methodology employed in this study is designed to extract meaningful trends from these datasets while accounting for their differences across both space and time. The intersection of annual composite images with the KBA shapefile allowed for extracting average night-time light values for each KBA for every year in the study period. This process creates a spatio-temporal dataset that captures how light levels within these ecologically significant areas have changed over nearly three decades [30]. The separation of the analysis into two distinct time periods (1992–2013 and 2014–2020) due to the use of different datasets allows for a more nuanced understanding of temporal trends. It enables the detection of potential shifts in trends that might coincide with the transition between datasets and the identification of more recent changes that might be particularly relevant to current conservation efforts. Spatial trends could be analyzed by mapping the categorized trends across KBAs, potentially revealing regional or global patterns in the relationship between night-time lighting changes and biodiversity hotspots. This spatial analysis could highlight areas where development pressures on KBAs are particularly intense or where conservation efforts might be having a positive impact in terms of limiting light pollution.

4.3.5 Validation of the Results

While not explicitly outlined in the original text, a robust validation and uncertainty assessment process would be crucial for ensuring the reliability of the results. We conducted a detailed cross-validation analysis for the overlapping year to address the transition between the DMSP OLS and VIIRS datasets in 2013–2014. We compared the light intensity values from both datasets for a large sample of pixels across different KBAs, calculating metrics such as correlation coefficients and mean absolute differences. This process helped us quantify the consistency between the two datasets and informed our interpretation of trends that spanned this transition period. To assess the classification accuracy of our trend categorizations (significant positive, positive non-significant, negative non-significant, and considerable negative), we manually interpreted the high-resolution images for a randomly selected subset of KBAs, comparing the observed changes in lighting infrastructure and urban development to our categorized trends. Each KBA in this subset was classified as correctly or incorrectly categorized based on the visual evidence of lighting changes.

4.3.6 Limitations and Considerations

While the methodology employed in this study is robust, it is essential to acknowledge its limitations and considerations. The most significant challenge lies in the transition between the DMSP OLS and VIIRS datasets in 2013–2014. Due to the different methodologies used in generating these datasets, direct comparisons between the two time periods should be made cautiously. Including background noise and transient events in the VIIRS data may affect the interpretation of trends, especially in areas with low light levels. The resolution differences between the datasets (30 arc seconds for DMSP OLS vs. 15 arc seconds for VIIRS) may also introduce some inconsistencies in the analysis, particularly for smaller KBAs or those with complex spatial development patterns.

4.4 Results and Discussion of Night-Time Lighting in Indian KBAs

4.4.1 Overall Trend Analysis

The analysis of night-time lighting data across KBAs in India reveals significant trends that provide crucial insights into the changing landscape of human activity and its potential impact on these important conservation sites. This study utilized two distinct datasets to capture a comprehensive picture of lighting changes over nearly three decades: stable night lights from the Defense Meteorological Satellite Program's Operational Linescan System (DMSP OLS) for the period 1992–2013 (Fig. 4.1) and average radiance data from the Visible Infrared Imaging Radiometer Suite (VIIRS) for 2014–2020 (Fig. 4.2). The results paint a stark picture of increasing artificial illumination across KBAs in India. For the earlier period (1992–2013), stable night lights showed a significant positive trend in 53.90% of the KBAs analyzed, indicating that over half of KBAs in India experienced substantial increases in artificial lighting over this two-decade span. The implications of this trend are profound, suggesting a widespread expansion of human activity and development in and around these ecologically crucial areas.

The findings from the more recent VIIRS data (2014–2020) are even more striking. During this period, 76.39% of KBAs exhibited a significant positive trend in average radiance. This accelerated change is alarming from a conservation perspective, indicating that three-quarters of KBAs in India are experiencing intensifying pressures from artificial lighting and associated human activities. The rapidity of this change suggests that the challenges facing biodiversity conservation in these areas may be escalating at an unprecedented rate. Notably, a small fraction of KBAs (2.71%) demonstrated a significant negative trend in average radiance during 2014–2020. While this percentage is small, it offers a glimmer of hope, showing that reductions in artificial lighting around sensitive areas are possible. These cases merit further investigation to understand the factors contributing to these positive outcomes, as they could provide valuable insights for conservation strategies elsewhere. The trend analysis underscores the pervasive nature of increasing night-time lighting across KBAs in India. This broad-scale change suggests that the pressures of human development and activity are not confined to a few isolated areas but are a widespread phenomenon affecting biodiversity hotspots across the country. The acceleration of this trend in recent years, as evidenced by the VIIRS data, is particularly concerning and calls for urgent attention from conservationists, policymakers, and land managers.

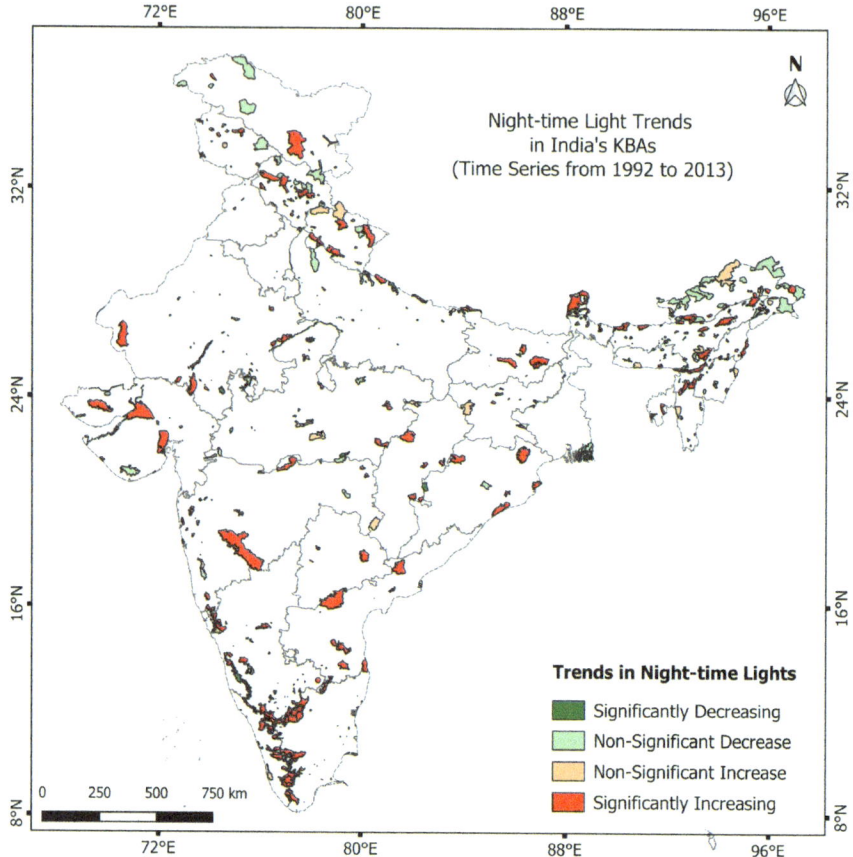

Fig. 4.1 Map visually represents the spatial distribution of night-time light trends across KBAs in India from 1992 to 2013, highlighting areas of concern and improvement in the effects of night-time lights

4.4.2 Regional Patterns

Examining the state-level patterns of night-time lighting changes reveals interesting regional variations and commonalities across India. These patterns provide valuable context for understanding the diverse pressures facing KBAs in different parts of the country and may help inform tailored conservation strategies. The trends are particularly prevalent in rapidly developing states such as Andhra Pradesh, Karnataka, and Tamil Nadu, showing significant positive trends across 80–95% of their KBAs. This high percentage likely reflects these regions' intense economic development and urbanization pace [31]. The challenge for these states lies in balancing their growth trajectories with preserving their rich biodiversity. Increasing night-time illumination is evident even in more remote and less densely populated areas. For instance,

4.4 Results and Discussion of Night-Time Lighting in Indian KBAs

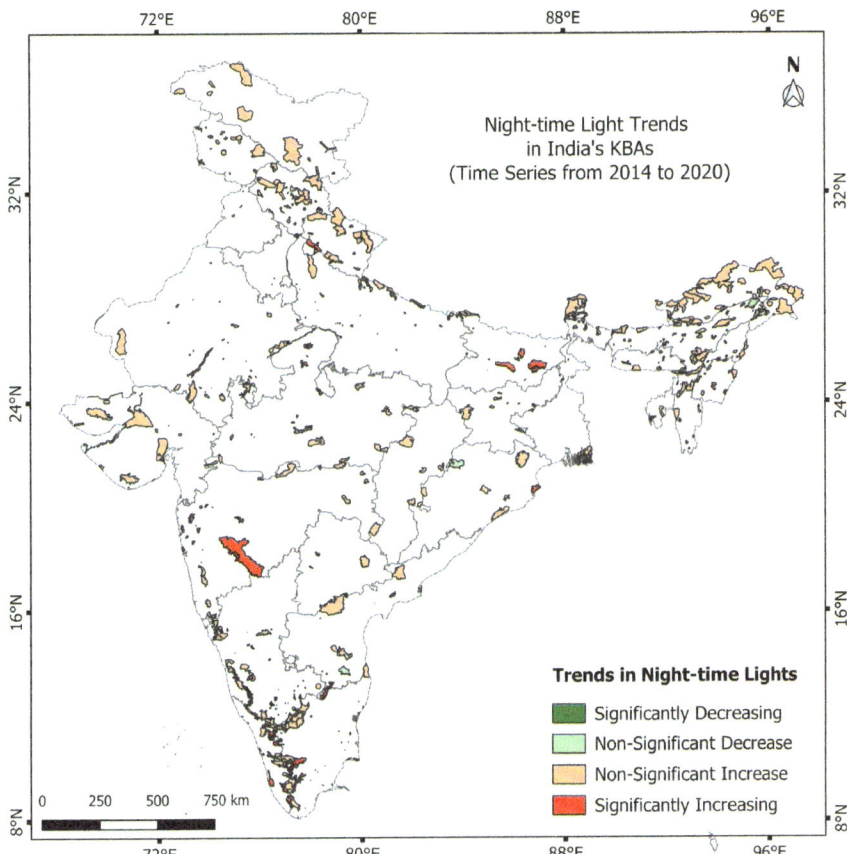

Fig. 4.2 Map visually represents the spatial distribution of night-time light trends across KBAs in India from 2014 to 2020, highlighting areas of concern and improvement in the effects of night-time lights

over 90% of KBAs exhibited increasing night-time lighting trends in Northeastern Arunachal Pradesh. This finding is noteworthy, given the state's reputation for pristine forests and rich biodiversity. It suggests that even areas traditionally considered less developed are not immune to the expanding reach of human activity and its associated lighting.

States with significant tourism industries, like Goa, present another exciting case. In Goa, over 80% of KBAs showed substantial increases in stable night lights from 1992 to 2013. This trend likely reflects the expansion of tourism infrastructure and associated development, highlighting the potential tensions between economic development through tourism and biodiversity conservation. Its presence in smaller states and union territories further emphasizes the trend. For example, all three KBAs in Tripura showed significant increases in stable night lights from 1992 to

2013. This ubiquity across diverse geographical and socioeconomic contexts underscores the widespread nature of this phenomenon and the need for a coordinated, national-level response. These regional patterns suggest that while the overall trend of increasing night-time illumination is consistent across India, this change's drivers and specific characteristics may vary significantly from one region to another. Understanding these regional nuances is crucial for developing effective, locally tailored conservation strategies.

4.4.3 Implications and Discussion

The widespread increase in night-time lighting across KBAs in India has far-reaching implications for biodiversity conservation. Artificial lighting can significantly disrupt the natural behavioral patterns of many species. Nocturnal animals, in particular, may find their foraging, mating, and migration patterns altered by artificial light. For example, sea turtle hatchlings, which naturally orient themselves toward the brighter horizon of the sea, can become disoriented by coastal lighting, leading them away from the ocean. Similarly, many bat species may avoid lit areas, potentially reducing their foraging range and efficiency. The increase in lighting around KBAs also signals potential edge effects. As human settlements and activities expand around the borders of protected areas, the quality of habitat, even within the KBA boundaries, can be degraded, leading to changes in species composition, with edge-tolerant species potentially outcompeting more sensitive interior species. The creation of these edge habitats can also facilitate the spread of invasive species, further threatening the integrity of the KBA ecosystems. Fragmentation is another serious concern arising from increased lighting. Brightly lit areas can act as barriers to movement for light-sensitive species, fragmenting populations. The fragmentation can reduce gene flow between populations, potentially decreasing genetic diversity and resilience. For example, some bat species have been observed to avoid crossing lit areas, which could limit their ability to access different parts of their habitat.

The alteration of species interactions is a subtler but potentially far-reaching impact of increased artificial lighting. Predator–prey dynamics can be significantly affected, as artificial lighting can increase prey species' vulnerability while potentially disorienting predators, leading to imbalances in ecosystem dynamics that may have cascading effects throughout the food web [32]. Physiological impacts on individual organisms are also a significant concern. Artificial lighting can disrupt these processes, affecting hormonal balances and immune function. For plants, artificial lighting can alter flowering times and influence interactions with pollinators, potentially leading to mismatches in plant–pollinator relationships. While some level of development around KBAs may be inevitable or beneficial for local communities, the widespread nature of these lighting increases suggests a need for more strategic land-use planning and lighting policies. The challenge lies in finding a balance between human needs and preserving biodiversity. The small percentage of KBAs (2.71%)

that showed significant decreases in average radiance from 2014 to 2020 offers an interesting counterpoint to the overall trend.

Further research into the policies or circumstances that enabled these reductions could provide valuable insights for conservation strategies elsewhere. The implications of these findings extend beyond the immediate impacts on biodiversity. Increased lighting often correlates with other human activity and development forms, such as road construction, urbanization, and industrial development. As such, the observed lighting trends may indicate broader human-caused pressures on these critical ecological areas.

4.4.4 Limitations and Future Research Directions

While this study provides valuable insights into the changing pressures on KBAs in India, it's essential to acknowledge its limitations and consider directions for future research. One fundamental limitation is that night-time lighting, while a helpful proxy, is an indirect measure of human activity and development. While increased lighting generally correlates with human presence and activity, the specific nature and intensity of that activity can vary widely. Future studies could benefit from ground-truthing and integrating complementary data on specific human activities to provide a more nuanced and complete picture of human-caused pressures on KBAs. Another limitation is that the impacts of increased lighting likely vary depending on the specific ecological characteristics and species present in each KBA. The sensitivity to light pollution can differ dramatically between species and ecosystems. Future research should investigate these site-specific effects, perhaps through case studies of particularly vulnerable or representative KBAs. While this study identified long-term trends, more detailed temporal analysis could reveal essential patterns in the pace and timing of lighting changes. For instance, investigating whether changes are gradual or occur in sudden jumps could provide insights into the nature of development processes affecting these areas.

Several promising directions for future research emerge from this study:

1. *Integration of Multiple Remote Sensing Indicators*: Combining night-time lighting data with other remote sensing indicators, such as forest cover change or urban expansion metrics, could provide a more comprehensive assessment of threats to KBAs [33]. This multifaceted approach could offer a more nuanced understanding of the interplay between different human-caused pressures.
2. *Policy Effectiveness Studies*: Investigating the effectiveness of dark sky policies and other lighting management strategies in reducing impacts on biodiversity in and around KBAs could provide valuable insights for policymakers and conservation managers [34].
3. *Species-Specific Impact Studies*: Conducting detailed, species-specific studies to understand better the ecological impacts of increased night-time lighting in

different KBA types would provide crucial information for targeted conservation efforts [35].
4. *Socioeconomic Analysis*: Exploring the socioeconomic drivers of lighting increases around KBAs could inform more effective conservation and development policies. Understanding the human dimensions of this issue is crucial for developing sustainable solutions that meet ecological and social needs [36].
5. *Technological Solutions*: Research into innovative lighting technologies and designs that minimize impacts on wildlife while meeting human needs could offer practical solutions to mitigate the effects of increased lighting [37].

4.5 Technology–Policy Interface to Limit the Spread of Stable Night Lights

The study's findings of increasing night-time illumination across KBAs in India call for an advanced technology–policy interface to address this concerning trend. At the forefront of this approach is the enhanced use of remote sensing and satellite technology [38]. Building on the study's utilization of DMSP OLS and VIIRS datasets, we propose leveraging high-resolution satellite imagery combined with advanced spectral analysis to provide more nuanced insights into light pollution patterns. Hyperspectral imaging satellites, for instance, could differentiate between various types of light sources, helping to identify the most ecologically disruptive emissions. The enhanced satellite monitoring should be mandated through policies requiring regular assessments of night light trends in and around KBAs, with the results indicating zoning regulations and development planning in sensitive regions.

Artificial intelligence and Machine learning technologies offer powerful tools for analyzing the vast amounts of data generated by these enhanced satellite observations and ground-based sensors [39]. Machine learning algorithms can be developed to predict future night light trends based on historical data and socioeconomic indicators, identify subtle patterns of light pollution that may impact specific species or ecosystems, and optimize adaptive lighting systems to minimize ecological impact while meeting human needs. To fully leverage these capabilities, policies should be implemented to support the development and deployment of these AI tools, including establishing AI research centers focused on ecological applications and integrating AI-driven insights into environmental impact assessments for new development projects near KBAs.

The insights gained from these advanced technologies should inform various policy applications. We recommend implementing AI-informed dynamic zoning regulations that adapt to changing light pollution patterns and ecological needs. Comprehensive light pollution impact assessments should be mandated for all significant development projects near KBAs, utilizing the latest remote sensing and AI analytics. Policies should also be developed to promote the adoption of intelligent, eco-friendly lighting technologies in both public and private sectors.

4.5 Technology–Policy Interface to Limit the Spread of Stable Night Lights

Ensuring effective implementation and compliance with light pollution policies is critical. We propose deploying AI-powered systems to continuously monitor compliance with lighting regulations using satellite and drone-gathered data. A blockchain-based system could be implemented to record compliance data, policy implementations, and their consequences, ensuring transparency and accountability [40]. This could be further enhanced by utilizing blockchain-based smart contracts to automatically enforce penalties or rewards based on compliance with lighting regulations. Policies should be established to recognize these technological enforcement mechanisms and to develop clear guidelines for their use, balancing efficiency with privacy and ethical considerations (Fig. 4.3).

Engaging the public can significantly enhance monitoring efforts and raise awareness about light pollution issues. We recommend developing mobile applications that allow citizens to report light pollution incidents in and around KBAs, participate in night sky brightness measurements using smartphone sensors, and access educational resources about light pollution and its ecological impacts. Policies should be implemented to officially recognize data collected through these citizen science initiatives and integrate this data into broader light pollution assessments and policy-making processes. Light pollution and climate change interface requires special attention [41–43]. We propose developing AI models integrating climate change projections with night light trend data to predict future impacts on KBAs. Policies for adaptive lighting management in and around KBAs should be implemented, considering potential shifts in species distributions due to climate change. Given

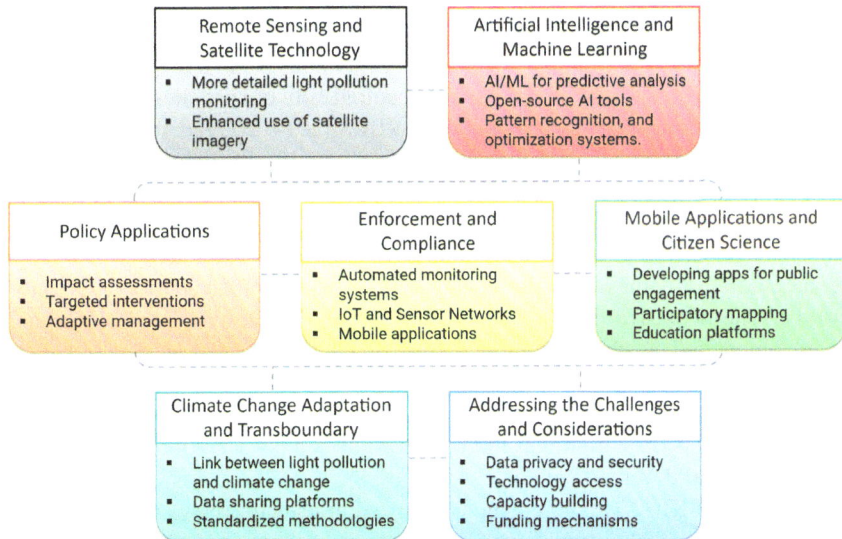

Fig. 4.3 Diagram outlines different technological tools and approaches that can be integrated to support the technology–policy interface to reduce the effects of night-time lights, from data analysis to policy implementation in KBAs

the transboundary nature of both light pollution and climate change impacts, it is crucial to establish international frameworks for sharing data and coordinating light pollution mitigation efforts.

In the implementation of this technology–policy interface, several issues must be addressed. Clear policies need to be established on collecting, using, and storing high-resolution night-time imagery, particularly in inhabited areas near KBAs, to ensure data privacy [44]. Programs should be developed to ensure that smaller municipalities and rural areas have access to advanced lighting technologies and monitoring capabilities, addressing issues of technological equity. Frameworks must be created to balance dark sky conservation with safety and economic development needs, potentially using AI to model optimal solutions [45]. Finally, policies should be designed to be flexible and adaptable, capable of adapting to technological advancements and new scientific understandings of light pollution. By addressing these challenges and leveraging the proposed technologies within a robust policy framework, India can work toward effectively managing night light pollution in its KBAs. This approach addresses the immediate concerns highlighted by the study and establishes a forward-looking system capable of adapting to future changes and challenges in biodiversity conservation. Through this comprehensive technology–policy interface, India can strive to preserve its crucial ecological areas' natural night-time environment while balancing human development needs.

4.6 Conclusion

The study on night-time lighting trends in KBAs in India reveals a concerning pattern of increasing artificial lighting across these ecologically crucial sites. By analyzing data from the DMSP OLS (1992–2013) and VIIRS (2014–2020) datasets, we found that most KBAs experienced significant increases in night-time lighting, accelerating in recent years. From 1992 to 2013, 53.90% of KBAs showed significant positive trends in stable night lights, while from 2014 to 2020, this proportion increased dramatically to 76.39% of KBAs exhibiting significant positive trends in average radiance. These findings have profound implications for biodiversity conservation in India. The pervasive increase in artificial lighting suggests widespread human encroachment and development in and around KBAs, potentially leading to habitat disturbance, fragmentation, and alterations in species behavior and interactions. The regional variations observed, with some states showing exceptionally high percentages of affected KBAs, highlight the need for tailored conservation strategies that account for local development contexts. While the overall trend is concerning, the small fraction of KBAs (2.71%) demonstrating significant decreases in average radiance from 2014 to 2020 offers hope. These cases merit further investigation as they may provide valuable insights into successful light pollution mitigation strategies.

The study underscores the urgent need for a robust technology–policy interface to address the spread of night-time lighting in KBAs. We recommend leveraging advanced remote sensing technologies, artificial intelligence, and machine learning to enhance monitoring capabilities and inform adaptive management strategies. Policy recommendations include implementing AI-informed dynamic zoning regulations, mandating comprehensive light pollution impact assessments for development projects, and establishing interstate and international agreements for managing transboundary light pollution. Future research directions should focus on integrating multiple remote sensing indicators, conducting species-specific impact studies, and exploring the socioeconomic drivers of lighting increases. Additionally, investigating the effectiveness of dark sky policies and developing innovative lighting technologies could provide practical solutions for mitigating the ecological impacts of artificial lighting.

In brief, this study serves as a call to action for policymakers, conservationists, and stakeholders to address the growing threat of light pollution to India's biodiversity. By implementing advanced technological solutions within a robust policy framework, India can work toward preserving the natural night-time environment of its KBAs while balancing human development needs. The challenge lies in fostering sustainable development practices that recognize the ecological value of darkness and prioritize the preservation of these critical habitats for future generations.

References

1. C.D. Elvidge, M. Zhizhin, T. Ghosh et al., Annual time series of global VIIRS nighttime lights derived from monthly averages: 2012 to 2019. Remote Sens. **13**, 922 (2021). https://doi.org/10.3390/rs13050922
2. C.D. Elvidge, K.E. Baugh, E.A. Kihn et al., Mapping city lights with nighttime data from the DMSP Operational Linescan System. Photogramm. Eng. Remote Sens. **63**, 727–734 (1997)
3. C. Small, F. Pozzi, C.D. Elvidge, Spatial analysis of global urban extent from DMSP-OLS night lights. Remote Sens. Environ. **96**, 277–291 (2005). https://doi.org/10.1016/J.RSE.2005.02.002
4. S. Rakkasagi, M.K. Goyal, S. Jha, Evaluating the future risk of coastal Ramsar wetlands in India to extreme rainfalls using fuzzy logic. J. Hydrol. **632**, 130869 (2024). https://doi.org/10.1016/j.jhydrol.2024.130869
5. S. Baidya, P. Chakraborty, S. Rakkasagi et al., Pathways to build resilience toward the impact of climate change on the Indian Sunderban, in *Ecosystem Restoration: Towards Sustainability and Resilient Development*. ed. by A.K. Gupta, M.K. Goyal, S.P. Singh (Springer Nature Singapore, Singapore, 2023), pp.307–333
6. C. Aubrecht, M. Jaiteh, A. de Sherbinin, *Global Assessment of Light Pollution Impact on Protected Areas* (2010)
7. A. Sánchez de Miguel, J. Zamorano, J. Gómez Castaño, S. Pascual, Evolution of the energy consumed by street lighting in Spain estimated with DMSP-OLS data. J. Quant. Spectrosc. Radiat. Transf. **139**, 109–117 (2014). https://doi.org/10.1016/j.jqsrt.2013.11.017
8. H.L. Beyer, O. Venter, H.S. Grantham, J.E.M. Watson, Substantial losses in ecoregion intactness highlight urgency of globally coordinated action. Conserv. Lett. **13** (2020). https://doi.org/10.1111/conl.12692
9. M. Kumar Goyal, V. Poonia, V. Jain, Three decadal urban drought variability risk assessment for Indian smart cities. J. Hydrol. **625**, 130056 (2023). https://doi.org/10.1016/j.jhydrol.2023.130056
10. V. Poonia, M. Kumar Goyal, S. Jha, S. Dubey, Terrestrial ecosystem response to flash droughts over India. J. Hydrol. **605**, 127402 (2022). https://doi.org/10.1016/j.jhydrol.2021.127402

11. K. Huang, X. Li, X. Liu, K.C. Seto, Projecting global urban land expansion and heat island intensification through 2050. Environ. Res. Lett. **14**, 114037 (2019). https://doi.org/10.1088/1748-9326/ab4b71
12. K.J. Gaston, T.W. Davies, S.L. Nedelec, L.A. Holt, Impacts of artificial light at night on biological timings. Annu. Rev. Ecol. Evol. Syst. **48**, 49–68 (2017). https://doi.org/10.1146/annurev-ecolsys-110316-022745
13. K.J. Gaston, J.P. Duffy, S. Gaston et al., Human alteration of natural light cycles: causes and ecological consequences. Oecologia **176**, 917–931 (2014). https://doi.org/10.1007/s00442-014-3088-2
14. D.M. Dominoni, J.C. Borniger, R.J. Nelson, Light at night, clocks and health: from humans to wild organisms. Biol. Lett. **12**, 20160015 (2016). https://doi.org/10.1098/rsbl.2016.0015
15. D. Sanders, E. Frago, R. Kehoe et al., A meta-analysis of biological impacts of artificial light at night. Nat. Ecol. Evol. **5**, 74–81 (2020). https://doi.org/10.1038/s41559-020-01322-x
16. B. Seymoure, Enlightening butterfly conservation efforts: the importance of natural lighting for butterfly behavioral ecology and conservation. Insects **9**, 22 (2018). https://doi.org/10.3390/insects9010022
17. V. Jain, K.S. Rautela, M.K. Goyal, *Ecological Restoration: An Overview of Science and Policy Regime* (2023), pp. 1–27
18. K. Jung, C.G. Threlfall, Urbanisation and its effects on bats—a global meta-analysis, in *Bats in the Anthropocene: Conservation of Bats in a Changing World*. (Springer International Publishing, Cham, 2016), pp.13–33
19. D. Karnad, K. Isvaran, C.S. Kar, K. Shanker, Lighting the way: towards reducing misorientation of Olive Ridley hatchlings due to artificial lighting at Rushikulya, India. Biol. Conserv. **142**, 2083–2088 (2009). https://doi.org/10.1016/j.biocon.2009.04.004
20. B. Kempenaers, P. Borgström, P. Loës et al., Artificial night lighting affects dawn song, extra-pair siring success, and lay date in songbirds. Curr. Biol. **20**, 1735–1739 (2010). https://doi.org/10.1016/j.cub.2010.08.028
21. X. Hu, Y. Qian, S.T.A. Pickett, W. Zhou, Urban mapping needs up-to-date approaches to provide diverse perspectives of current urbanization: a novel attempt to map urban areas with nighttime light data. Landsc. Urban Plan. **195**, 103709 (2020). https://doi.org/10.1016/j.landurbplan.2019.103709
22. N. Kumar, M.K. Goyal, Projected changes in monsoonal compound dry-hot extremes in India. Atmos. Res. **310**, 107605 (2024). https://doi.org/10.1016/j.atmosres.2024.107605
23. K.S. Rautela, S. Singh, M.K. Goyal, Aerosol atmospheric rivers: patterns, impacts, and societal insights. Environ. Sci. Pollut. Res. (2024). https://doi.org/10.1007/s11356-024-34625-8
24. K.S. Rautela, S. Singh, M.K. Goyal, Resilience to air pollution: a novel approach for detecting and predicting aerosol atmospheric rivers within earth system boundaries. Earth Syst. Environ. (2024). https://doi.org/10.1007/s41748-024-00421-0
25. N. Kumar, V. Poonia, B.B. Gupta, M.K. Goyal, A novel framework for risk assessment and resilience of critical infrastructure towards climate change. Technol. Forecast. Soc. Change **165**, 120532 (2021). https://doi.org/10.1016/j.techfore.2020.120532
26. S. Rakkasagi, M.K. Goyal, Assessing risk levels of the extreme rainfalls in Ramsar wetlands of India using fuzzy logic, in *Chapman Conference on Remote Sensing of the Water Cycle* (AGU, 2024)
27. M.K. Goyal, S. Rakkasagi, S. Shaga et al., Spatiotemporal-based automated inundation mapping of Ramsar wetlands using Google Earth Engine. Sci. Rep. **13**, 17324 (2023). https://doi.org/10.1038/s41598-023-43910-4
28. S. Singh, A. Kumar, Understanding the intricacies of rainfall dynamics using entropy measures. J. Water Clim. Change (2024). https://doi.org/10.2166/wcc.2024.350
29. S. Singh, D. Kumar, D.K. Vishwakarma et al., Seasonal rainfall pattern using coupled neural network-wavelet technique of southern Uttarakhand, India. Theor. Appl. Climatol. **155**, 5185–5201 (2024). https://doi.org/10.1007/s00704-024-04940-8
30. V. Gupta, S. Rakkasagi, S. Rajpoot et al., Spatiotemporal analysis of Imja Lake to estimate the downstream flood hazard using the SHIVEK approach. Acta Geophys. (2023). https://doi.org/10.1007/s11600-023-01124-2

References

31. S. Bhardwaj, P. Machiwar, C. Kant et al., *Analysis of Urbanization and Assessment of Its Impact on Groundwater and Land Use/Land Cover Using GIS Techniques: A Case Study of Bhopal and Gurugram District* (2023), pp. 219–255
32. P. Patle, P.K. Singh, S. Rakkasagi et al., *Application of Water Accounting Plus Framework for the Assessment of the Water Consumption Pattern and Food Security* (2023), pp. 257–269
33. A.E. Beresford, P.F. Donald, G.M. Buchanan, Repeatable and standardised monitoring of threats to Key Biodiversity Areas in Africa using Google Earth Engine. Ecol. Indic. **109**, 105763 (2020). https://doi.org/10.1016/j.ecolind.2019.105763
34. D.A. Silver, G.M. Hickey, Managing light pollution through dark sky areas: learning from the world's first dark sky preserve. J. Environ. Plan. Manag. **63**, 2627–2645 (2020). https://doi.org/10.1080/09640568.2020.1742675
35. K.J. Gaston, J. Bennie, T.W. Davies, J. Hopkins, The ecological impacts of nighttime light pollution: a mechanistic appraisal. Biol. Rev. **88**, 912–927 (2013). https://doi.org/10.1111/brv.12036
36. S.L. Maxwell, V. Cazalis, N. Dudley et al., Area-based conservation in the twenty-first century. Nature **586**, 217–227 (2020). https://doi.org/10.1038/s41586-020-2773-z
37. F. Falchi, P. Cinzano, C.D. Elvidge et al., Limiting the impact of light pollution on human health, environment and stellar visibility. J. Environ. Manage. **92**, 2714–2722 (2011). https://doi.org/10.1016/j.jenvman.2011.06.029
38. S.R. Subramoniam, S. Ravindranath, S. Rakkasagi, H. Ram, *Water Resource Management Studies at Micro Level Using Geospatial Technologies* (2022), pp. 49–74
39. Y. Xu, X. Liu, X. Cao et al., Artificial intelligence: a powerful paradigm for scientific research. Innovation **2**, 100179 (2021). https://doi.org/10.1016/j.xinn.2021.100179
40. M. Regona, T. Yigitcanlar, C. Hon, M. Teo, Artificial intelligence and sustainable development goals: systematic literature review of the construction industry. Sustain. Cities Soc. **108**, 105499 (2024). https://doi.org/10.1016/j.scs.2024.105499
41. M.K. Goyal, A.K. Gupta, S. Jha et al., Climate change impact on precipitation extremes over Indian cities: non-stationary analysis. Technol. Forecast. Soc. Change **180**, 121685 (2022). https://doi.org/10.1016/j.techfore.2022.121685
42. M.K. Goyal, A.K. Gupta, J. Das et al., Heatwave magnitude impact over Indian cities: CMIP 6 projections. Theor. Appl. Climatol. **154**, 959–971 (2023). https://doi.org/10.1007/s00704-023-04599-7
43. S. Rakkasagi, V. Poonia, M.K. Goyal, Flash drought as a new climate threat: drought indices, insights from a study in India and implications for future research. J. Water Clim. Change (2023). https://doi.org/10.2166/wcc.2023.347
44. B. Yu, F. Chen, C. Ye et al., Temporal expansion of the nighttime light images of SDGSAT-1 satellite in illuminating ground object extraction by joint observation of NPP-VIIRS and sentinel-2A images. Remote Sens. Environ. **295**, 113691 (2023). https://doi.org/10.1016/j.rse.2023.113691
45. U.A.K. Betz, L. Arora, R.A. Assal et al., Game changers in science and technology—now and beyond. Technol. Forecast. Soc. Change **193**, 122588 (2023). https://doi.org/10.1016/j.techfore.2023.122588

Chapter 5
Policy Recommendations and Future Directions

5.1 Achievement of the SDGs

This section explores the critical role of Key Biodiversity Areas (KBAs) in India in achieving the sustainable development goals (SDGs; Fig. 5.1). By examining various aspects of KBA conservation, including forest tree cover changes, forest fire dynamics, and night-time lighting trends, the section provides valuable insights into the challenges and opportunities for sustainable development in India. The study presented in this chapter contributes significantly to our understanding of how protecting and managing KBAs can support multiple SDGs, particularly those related to biodiversity conservation, climate action, and sustainable land use.

5.1.1 SDGs for Forest Cover Change

The previous Chap. 2 analyzed forest cover loss trends across KBAs in India from 2001 to 2020, emphasizing the role of the technology–policy interface in reducing deforestation. The study utilizes Google Earth Engine [1, 2] to analyze 615 KBAs, providing crucial insights into conservation effectiveness and informing policy decisions. The findings reveal a complex picture of forest cover change:

- 32.05% of KBAs showed statistically significant decreases in forest loss rates, indicating successful conservation efforts in these areas.
- 10.36% of KBAs showed substantial increases in forest loss rates, highlighting areas of concern requiring immediate attention.
- Northeastern states emerge as deforestation hotspots, with Manipur showing 57.14% of KBAs experiencing increasing forest cover loss.
- Central Indian states like Chhattisgarh exhibit conservation success, with 77.78% of KBAs indicating significant declines in deforestation rates.

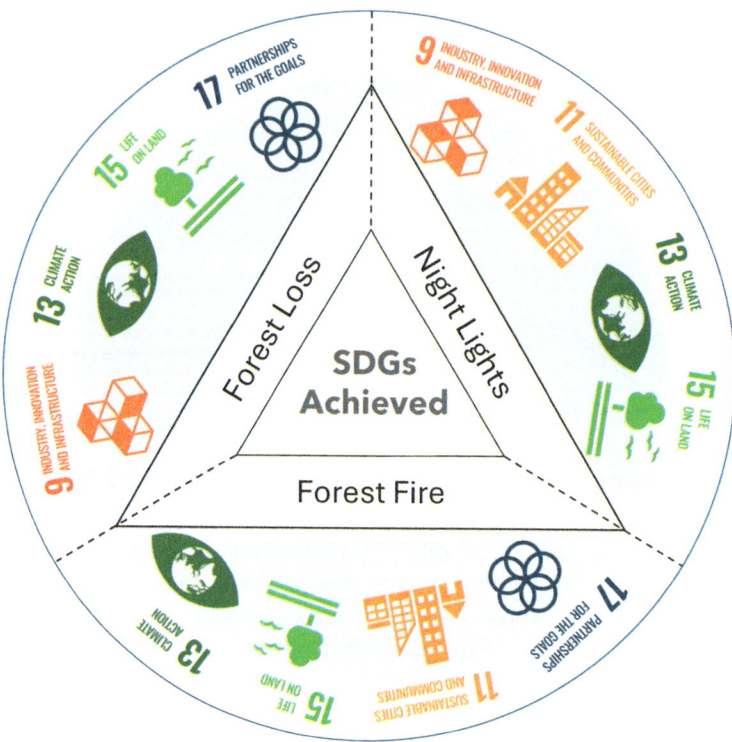

Fig. 5.1 Map diagram illustrates how the study results on three threats (i.e., forest loss, forest fire, and night-time lights) relate to various SDG achievements. The central triangle highlights three threats that suggest their interrelation in the context of SDG achievement

These results have significant implications for achieving multiple SDGs:

1. SDG 15 (Life on Land): The analysis directly contributes to this goal by providing crucial data on the state of terrestrial ecosystems. The finding that 32.05% of KBAs showed significant decreases in forest loss rates indicates progress toward halting deforestation and protecting biodiversity. However, the increasing forest loss in 10.36% of KBAs highlights the need for targeted interventions in these areas to preserve critical habitats and species.
2. SDG 13 (Climate Action): The study supports climate action efforts by monitoring forest cover changes in KBAs. Forests act as important carbon sinks, and understanding deforestation patterns helps develop strategies to mitigate climate change. The success in reducing forest loss in central Indian states provides valuable lessons for climate change mitigation through forest conservation.
3. SDG 17 (Partnerships for the Goals): The study emphasizes the need for transboundary cooperation in managing shared ecosystems, particularly in the Northeast and Himalayas, aligning with SDG 17's focus on strengthening global partnerships for sustainable development.

4. SDG 9 (Industry, Innovation, and Infrastructure): The chapter highlights advanced technologies like remote sensing and data analytics in monitoring forest cover changes. This innovation in conservation practices supports SDG 9's aim to foster innovation and build resilient infrastructure.

We conclude by emphasizing the need for targeted, context-specific interventions to address regional disparities in forest cover change. This approach supports the achievement of multiple SDGs by promoting more effective and efficient conservation strategies.

5.1.2 SDGs for Forest Fire Trends

The previous Chap. 3 analyzed forest fire trends in KBAs across India from 2001 to 2020 using satellite-derived fire data. The study aims to evaluate spatial and temporal patterns of fire events in these vital biodiversity areas and assess the implications for conservation and management.

Using the Fire Information for Resource Management System (FIRMS) dataset through Google Earth Engine, the study comprehensively analyzes fire trends across all designated KBAs in India. The results reveal a complex picture:

1. 6.70% of KBAs showed statistically significant increases in fire events.
2. 31.25% of KBAs exhibited significant decreasing trends in fire occurrences.
3. Notable regional variations were observed, with Northeastern states and parts of the Himalayas showing mainly concerning upward trends in fire events.

These findings have significant implications for achieving multiple SDGs:

1. SDG 15 (Life on Land): By analyzing fire trends in KBAs, the study directly contributes to SDG 15's target of protecting and restoring terrestrial ecosystems [3]. The finding that 31.25% of KBAs showed significant decreasing trends in fire events indicates progress in managing and reducing the impact of fires on biodiversity. However, the increasing fire trends in 6.70% of KBAs highlight areas requiring urgent intervention to protect critical habitats and species.
2. SDG 13 (Climate Action): Forest fires contribute to greenhouse gas emissions and can exacerbate climate change. The study supports efforts to mitigate climate change impacts by monitoring and analyzing fire trends, aligning with SDG 13 [4]. Identifying fire hotspots, particularly in the Northeast and Himalayas, allows for targeted climate action strategies in these vulnerable regions.
3. SDG 11 (Sustainable Cities and Communities): The study's focus on fire management in KBAs, especially those near urban areas, contributes to building resilient communities and reducing the risk of disasters, supporting SDG 11. Understanding fire trends can inform urban planning and disaster preparedness strategies [5].

4. SDG 17 (Partnerships for the Goals): The study highlights the need for international cooperation in fire management, particularly for KBAs that span international boundaries, aligning with SDG 17's emphasis on strengthening global partnerships.

Our recommendations for integrating advanced technologies with robust policy frameworks for fire management support the achievement of multiple SDGs by promoting more effective and sustainable conservation practices.

5.1.3 SDGs for Night-Time Lighting Trends

The previous Chap. 4 analyzed the trends in night-time lighting across KBAs in India, using two datasets: the DMSP OLS for 1992–2013 and the VIIRS for 2014–2020. The study aims to assess the extent of artificial lighting increase in KBAs and its potential effects on biodiversity conservation.

The findings reveal a concerning trend of increasing artificial illumination across KBAs:

1. 53.90% of KBAs showed significant positive trends in stable night lights from 1992 to 2013.
2. This trend accelerated recently, with 76.39% of KBAs showing significant positive trends in average radiance from 2014 to 2020.
3. Regional analysis showed variations across states of India, with some displaying exceptionally high percentages of affected KBAs.
4. Only 2.71% of KBAs showed significant decreases in average radiance from 2014 to 2020.

These results have significant implications for achieving multiple SDGs:

1. SDG 15 (Life on Land): The analysis of night-time lighting trends in KBAs directly contributes to SDG 15 by providing crucial data on human encroachment and development pressures on biodiversity hotspots. The widespread increase in artificial lighting indicates habitat disruption and fragmentation, potentially affecting species' behavior and interactions, highlighting the urgent need for conservation measures to protect terrestrial ecosystems.
2. SDG 11 (Sustainable Cities and Communities): By examining the spread of artificial lighting around KBAs, the study contributes to understanding the impact of urbanization on natural habitats. This information is crucial for developing sustainable urban planning strategies that minimize ecological disruption, supporting SDG 11 [6]. The study emphasizes the need for more strategic land-use planning and lighting policies that balance human needs with biodiversity conservation.
3. SDG 13 (Climate Action): The study's recommendation to develop AI models integrating climate change projections with night light trend data supports SDG

13 by enhancing our ability to predict and mitigate the combined impacts of climate change and light pollution on biodiversity [7].
4. SDG 9 (Industry, Innovation, and Infrastructure): The chapter proposes leveraging advanced technologies such as AI, machine learning, and blockchain for monitoring and managing light pollution. This technological innovation aligns with SDG 9's focus on fostering innovation and building resilient infrastructure.

Our recommendations for implementing a comprehensive technology–policy interface to address light pollution support the achievement of multiple SDGs by promoting more effective and sustainable conservation practices while balancing human development needs.

5.2 Future Technological Advances in KBA Monitoring

The need for advanced monitoring technologies becomes paramount as the KBA sites face increasing threats from human activities and climate change [8, 9]. This section explores future technological advances in KBA monitoring, focusing on how these innovations can enhance our ability to protect and manage these vital ecosystems (Fig. 5.2). The future of KBA monitoring will likely see significant advancements in remote sensing and satellite technology. Building upon current systems like MODIS, VIIRS, and high-resolution commercial satellites [10, 11], future satellites are expected to offer even greater spatial and temporal resolution [12, 13]. These improvements will allow for more detailed and frequent observations of KBAs, enabling near-real-time monitoring of changes in forest cover, fire occurrences, and night-time lighting. Hyperspectral imaging satellites represent an auspicious advancement.

Hyperspectral imagers can collect data across hundreds of narrow spectral bands, unlike current multispectral sensors [14]. This capability will allow for a more nuanced analysis of ecosystem health, enabling researchers to detect subtle changes in vegetation composition, identify specific plant species, and assess plant stress levels. In the context of KBA monitoring, hyperspectral imaging could provide early warning signs of ecosystem degradation or invasive species spread. Another exciting development is the potential for geostationary satellites dedicated to environmental monitoring. While most Earth observation satellites are in low Earth orbit and only pass over a given area periodically, geostationary satellites remain fixed relative to the Earth's surface. The geostationary satellites could enable continuous monitoring of KBAs, providing unprecedented temporal resolution for detecting rapid changes such as fires or illegal logging activities. UAVs and drones also play an increasingly important role in KBA monitoring. Future advancements in drone technology are likely to include longer flight times, greater payload capacity, and enhanced sensors [15]. These improvements will allow for more comprehensive surveys of KBAs, especially in remote or inaccessible areas. One potential innovation is the development of autonomous drone swarms for KBA monitoring. These swarms

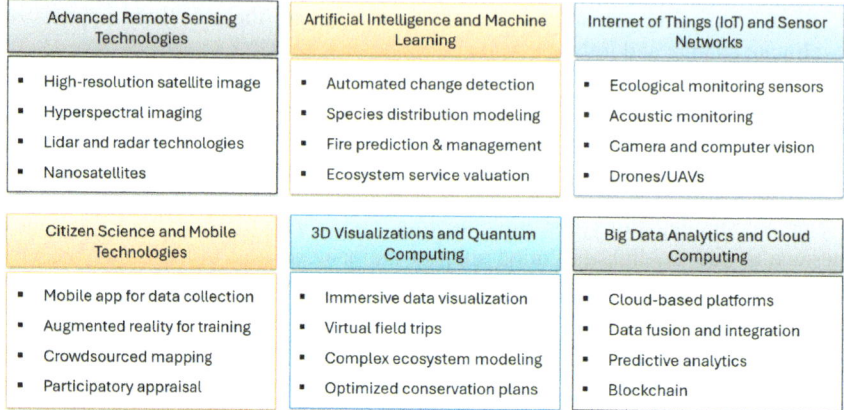

Fig. 5.2 Emerging technologies for KBAs conservation and ecosystem management highlight the diverse technological innovations being leveraged to enhance biodiversity monitoring, ecosystem assessment, and conservation planning

could work collaboratively to survey large areas quickly and efficiently, sharing real-time data and adapting their flight patterns based on observations. Equipped with various sensors, including high-resolution cameras, thermal imagers, and even chemical sensors, these drone swarms could provide a holistic view of ecosystem health and detect a wide range of potential threats. Another exciting possibility is integrating AI-powered image recognition systems into drones to identify specific species, detect signs of ecosystem disturbance, or spot illegal activities in real-time during flights.

Similarly, the role of artificial intelligence (AI) and machine learning (ML) in KBA monitoring will grow significantly in the coming years [16]. These technologies will be crucial in processing and analyzing the vast data generated by advanced remote sensing systems and ground-based sensors. One area where AI is likely to have a significant impact is predictive modeling [17]. By analyzing historical data on forest cover changes, fire occurrences, night-time lighting trends, and other relevant factors such as climate data and human development patterns, AI systems could predict future threats to KBAs with increasing accuracy, targeting resources to areas at the highest risk [18]. Machine learning algorithms can be trained to automatically detect and classify satellite or drone imagery features, such as different vegetation types, signs of deforestation, or evidence of human encroachment [19, 20]. As these algorithms improve, they will enable faster and more accurate analysis of large-scale imagery datasets, providing near-real-time insights into the state of KBAs. Another promising application of AI is in the field of bioacoustics monitoring. AI algorithms can identify and classify animal calls from audio recordings, allowing for non-invasive biodiversity tracking within KBAs. Future systems might combine widespread deployment of low-cost acoustic sensors with cloud-based AI analysis, providing continuous monitoring of animal populations and early warning

5.2 Future Technological Advances in KBA Monitoring

of potential ecosystem changes. The Internet of Things (IoT) holds great potential for enhancing KBA monitoring. In the future, we can expect to see widespread deployment of interconnected sensor networks throughout KBAs. These networks could include various sensor types, such as weather stations, air and water quality monitors, soil moisture sensors, and wildlife tracking devices [21]. One exciting possibility is the development of biodegradable IoT sensors. Such sensors might monitor soil conditions, detect the presence of specific species, or alert them to potential threats like fires or pollution events [22, 23]. Advanced wildlife tracking systems represent another promising application of IoT technology. Future tracking devices could be smaller, longer lasting, and capable of collecting more data. These devices transmit data in real-time to satellite networks, providing unprecedented insights into animal movements, behaviors, and responses to environmental changes within KBAs. The integration of edge computing with IoT sensor networks is another potential advancement. By processing data at the sensor level, these systems could reduce data transmission needs and provide faster alerts. For example, a network of edge-enabled camera traps could perform on-device image recognition, only alerting central systems when specific species of interest are detected.

While not directly a monitoring technology, blockchain, and other distributed ledger technologies have the potential to revolutionize data management and verification in KBA monitoring. These technologies could provide a secure, transparent, and tamper-proof monitoring data record, ensuring long-term datasets' integrity. Blockchain could also facilitate the creation of decentralized autonomous organizations for KBA management. These digital entities could automatically allocate resources for conservation efforts based on real-time monitoring data, ensuring rapid and efficient responses to emerging threats [24]. Smart contracts built on blockchain platforms could automate the enforcement of conservation agreements. For example, payments for ecosystem services could be automatically generated based on verified monitoring data, incentivizing local communities to protect KBAs. Likewise, the virtual reality (VR) and augmented reality (AR) technologies have the potential to transform how we visualize and interact with KBA monitoring data. Future systems might allow conservationists and policymakers to immerse themselves in virtual representations of KBAs, visualizing changes over time and exploring future scenarios based on predictive models. AR systems could enhance field monitoring efforts by overlaying real-time data and historical information onto a researcher's view of a KBA, aiding in species identification, highlighting areas of concern, and providing instant access to relevant data in the field. While still in its early stages, quantum computing has the potential to revolutionize data analysis for KBA monitoring in the long term. Quantum computers could process vast amounts of complex environmental data far more quickly than classical computers, enabling more sophisticated ecosystem models and predictive analyses. Quantum machine learning algorithms could discover subtle patterns in monitoring data that are beyond the capabilities of classical systems, potentially revealing new insights into ecosystem dynamics and improving our ability to predict and mitigate threats to KBAs.

5.3 Policy Recommendations for KBA Protection

KBAs are globally significant sites for biodiversity conservation, identified using standardized criteria formed by the International Union for Conservation of Nature (IUCN) [25]. As critical habitats for threatened species and ecosystems, KBAs play a vital role in achieving global biodiversity targets and sustainable development goals. However, these areas face increasing pressures from human activities, including deforestation, forest fires, and light pollution. A comprehensive set of policy recommendations is needed to address these challenges and enhance the protection of KBAs in India (Fig. 5.3). These recommendations should leverage advanced technologies, promote adaptive management strategies, and foster collaboration across various stakeholders.

One of the fundamental requirements for adequate KBA protection is establishing robust, integrated monitoring and assessment systems. These systems should leverage advanced remote sensing technologies, artificial intelligence (AI), and machine learning (ML) to provide real-time, high-resolution data on various ecological indicators [26]. Policies should mandate using multisensor satellite imagery to regularly monitor KBAs, including forest cover, fire occurrences, and nighttime lighting. Developing and implementing AI and ML algorithms to analyze

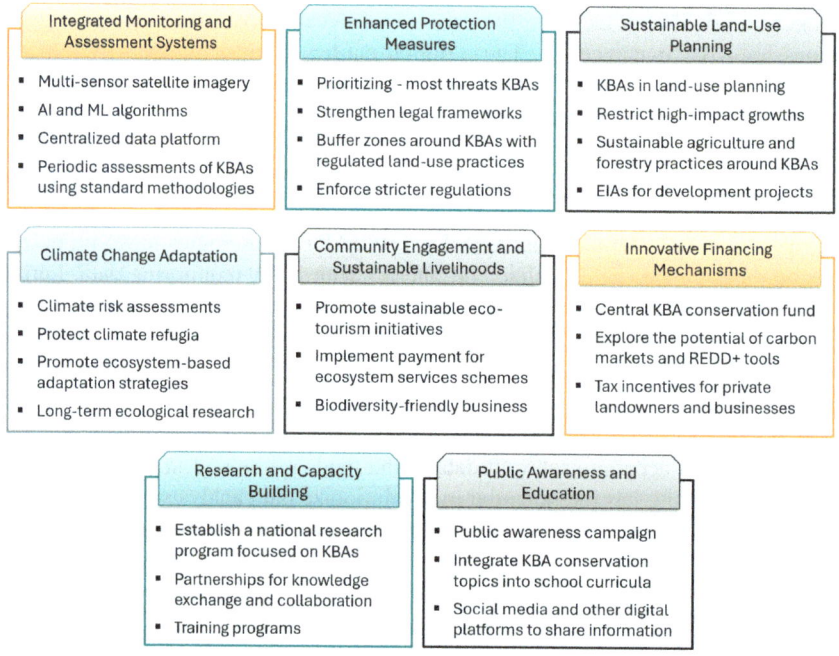

Fig. 5.3 Diagram outlines various interconnected policy areas crucial for effective KBA conservation, illustrating a holistic approach to KBA protection that combines technological, legal, economic, and social interventions

5.3 Policy Recommendations for KBA Protection

satellite data, predict trends, and identify areas of concern proactively is crucial. Establishing a centralized data platform that integrates information from various sources, including satellite imagery, ground-based sensors, and citizen science initiatives, would enhance monitoring capabilities. Regular comprehensive assessments of KBAs using standardized methodologies should be required to track changes in biodiversity, ecosystem health, and human pressures. Additionally, investing in research and development of new monitoring technologies, such as hyperspectral imaging and LiDAR, can enhance the accuracy and detail of ecological assessments [27]. Given the dynamic nature of ecosystems and the evolving pressures they face, KBA protection policies should incorporate adaptive management principles. This approach allows for continuous learning and adjustment of conservation strategies based on monitoring results and new scientific understanding. Policies should mandate developing and regularly updating adaptive management plans for each KBA, incorporating the latest monitoring data and scientific findings. Establishing clear thresholds for key ecological indicators that trigger specific management responses when exceeded is essential. A periodic review and review of KBA boundaries and management strategies must be undertaken to reflect the changes in ecological conditions and species distributions. Flexible funding mechanisms are required to quickly allocate resources to address emerging threats or capitalize on conservation opportunities. Promoting experimental approaches to conservation, encouraging controlled trials of different management strategies, and rigorous evaluation of their effectiveness can lead to more effective protection measures.

While many KBAs are already under some form of legal protection, the increasing pressures they face call for strengthened and more comprehensive protection measures. Policies should aim to expand the coverage of protected areas to encompass a more significant proportion of KBAs, prioritizing those facing the most critical threats. Strengthening legal frameworks for KBA protection, including increased penalties for illegal activities and enhanced enforcement capabilities, is crucial [28]. Implementing buffer zones around KBAs with regulated development and land-use practices can help minimize edge effects and maintain ecosystem connectivity. Stricter regulations on light pollution in and around KBAs should be developed and enforced, including requirements for dark sky-friendly lighting technologies. Comprehensive fire management plans for fire-prone KBAs, including prevention strategies, early detection systems, and rapid response protocols, are essential for protecting these vital areas. Many threats to KBAs stem from unsustainable land-use practices in the surrounding regions. Integrating KBA protection into broader land-use planning processes is crucial for long-term conservation success. Policies should mandate the inclusion of KBA conservation objectives in regional and national land-use planning processes. Developing and implementing zoning regulations that restrict high-impact development activities in areas adjacent to KBAs is necessary. Promoting sustainable agriculture and forestry practices in buffer zones around KBAs can help reduce pressure on protected ecosystems [29]. Policies incentivizing the restoration of degraded lands outside of KBAs can further reduce pressure on intact ecosystems [30]. Requiring comprehensive environmental impact assessments for all significant development projects near KBAs, with stringent mitigation measures for potential adverse impacts, is essential for safeguarding these critical areas.

Climate change significantly threatens many biodiversity areas, potentially altering species distributions and ecosystem dynamics [31]. Policies for KBA protection must incorporate climate change considerations to ensure long-term effectiveness. Developing climate vulnerability assessments for all KBAs and integrating findings into management plans is crucial. Identifying and protecting climate refugia within and around KBAs can provide safe havens for species as climates change. Policies should aim to maintain and enhance ecosystem connectivity, facilitating species movement in response to changing climatic conditions. Promoting ecosystem-based adaptation strategies can improve the resilience of KBAs to climate impacts while providing co-benefits for biodiversity and local communities. Establishing long-term ecological research sites within KBAs to monitor and study the effects of climate change on biodiversity and ecosystem processes can provide valuable insights for adaptive management. Adequate KBA protection requires the support and involvement of local communities. Policies should aim to align conservation goals with sustainable livelihood opportunities for communities living in and around KBAs. Developing and implementing community-based natural resource management programs that give local communities a stake in KBA conservation is essential. Promoting sustainable eco-tourism initiatives can generate income for local communities while minimizing ecological impacts on KBAs. Implementing payment for ecosystem services (PES) schemes can compensate local communities for conservation activities and sustainable land management practices [32]. Supporting the development of sustainable, biodiversity-friendly enterprises can provide alternative livelihoods and reduce pressure on KBA resources. Ensuring indigenous and local communities' rights and traditional knowledge are respected and integrated into KBA management strategies is crucial for successful conservation efforts.

Many KBAs, particularly in regions like the Northeast and Himalayas, are part of larger transboundary ecosystems. Adequate protection of these areas requires international cooperation. Policies should establish bilateral and multilateral agreements for the joint management and monitoring of transboundary KBAs. Developing shared databases and monitoring protocols for transboundary ecosystems can ensure consistent and comparable data collection. Implementing coordinated conservation strategies across national borders, including joint patrols, synchronized management interventions, and shared research initiatives, is essential. Creating mechanisms for rapid information sharing and coordinated response to transboundary threats such as fires, poaching, or disease outbreaks can enhance protection efforts [33]. Establishing transboundary protected areas or peace parks encompassing KBAs spanning national borders can foster international cooperation in conservation. Adequate funding is crucial for the long-term protection of KBAs. Innovative financing mechanisms can complement traditional government and donor funding sources. Policies should establish a national KBA conservation fund supported by government allocations, international donors, and private sector contributions. Implementing biodiversity offsetting policies can require developers to invest in KBA protection and restoration to compensate for biodiversity losses elsewhere. Developing green bonds and

5.3 Policy Recommendations for KBA Protection

other financial instruments to fund KBA conservation and sustainable development projects can attract new funding sources. Exploring the potential of carbon markets and REDD+ mechanisms to generate funding for forest conservation in KBAs can provide additional resources. Creating tax incentives for private landowners and businesses to support KBA conservation efforts can encourage broader participation in protection initiatives.

Adequate KBA protection requires a solid scientific foundation and skilled personnel. Policies should promote research and capacity building to enhance the knowledge base and implementation capabilities. Establishing a national research program focused on KBA ecology, conservation biology, and sustainable management practices can drive innovation in protection strategies. Developing partnerships between research institutions, conservation organizations, and government agencies can facilitate knowledge exchange and collaborative research. Implementing training programs for protected area managers, rangers, and local community members on advanced monitoring technologies, data analysis, and adaptive management techniques is crucial for effective implementation. Creating a scholarship program to support students and early career researchers working on KBA-related topics can foster the next generation of conservation leaders. Establishing a network of field stations within KBAs can facilitate long-term ecological research and monitoring.

Building public support for KBA conservation is crucial for the long-term success of protection efforts. Policies should aim to raise awareness about the importance of KBAs and engage the broader public in conservation initiatives. Developing a national public awareness campaign highlighting the importance of KBAs and their role in providing ecosystem services and supporting sustainable development can foster public support [34]. Integrating KBA conservation topics into school curricula at various levels can promote environmental awareness from an early age. Establishing visitor centers and educational programs at selected KBAs can provide hands-on learning experiences for students and the public. Supporting citizen science initiatives that engage the public in KBA monitoring and research activities can promote active participation in conservation efforts. Leveraging social media and other digital platforms to share information about KBAs and conservation successes can foster the public ownership and stewardship [35].

Protecting KBAs is crucial for preserving India's rich natural heritage and contributing to global biodiversity conservation efforts. The policy recommendations outlined above provide a comprehensive framework for enhancing KBA protection, leveraging advanced technologies, promoting adaptive management, and fostering collaboration across various stakeholders. Implementing these policies will require significant political will, financial investment, and coordination across multiple sectors and levels of government. However, the potential benefits are substantial for biodiversity conservation and ecosystem services, climate change mitigation and adaptation, and sustainable development [36–38]. As global environmental pressures intensify, the importance of KBAs as reservoirs of biodiversity and providers of essential ecosystem services will only increase. By implementing robust, forward-looking policies for KBA protection now, India can safeguard these critical areas for future generations and set an example for effective biodiversity conservation in the face of complex, multifaceted challenges. The success of these policy recommendations will ultimately depend on their effective implementation, regular evaluation, and adaptive refinement based on monitoring results and new scientific understanding. By committing to a comprehensive, science-based approach to KBA protection, India can make significant strides in preserving its natural heritage while contributing to global efforts to halt biodiversity loss and promote sustainable development.

5.4 Conclusion

The comprehensive analysis of KBAs in India presented in this chapter underscores the critical importance of these areas in achieving multiple SDGs and preserving India's rich biodiversity. This study has revealed a complex landscape of conservation challenges and opportunities by examining trends in forest cover change, forest fire occurrences, and night-time lighting across KBAs from 2001 to 2020. The findings demonstrate that while significant progress has been made in some areas, with 32.05% of KBAs showing decreases in forest loss rates and 31.25% exhibiting declining trends in fire occurrences, there are still considerable challenges to address. The increasing forest loss in 10.36% of KBAs and the alarming rise in artificial lighting across 76.39% of KBAs in recent years highlight the ongoing threats to these critical ecosystems. These results have significant implications for multiple SDGs, particularly SDG 15 (Life on Land), SDG 13 (Climate Action), and SDG 11 (Sustainable Cities and Communities). The study's findings emphasize the interconnectedness of these goals and the potential for KBA conservation to contribute simultaneously to biodiversity protection, climate change mitigation, and sustainable urban development.

The chapter's exploration of future technological advances in KBA monitoring offers promising avenues for enhancing conservation efforts. Advancements in remote sensing, artificial intelligence, Internet of Things (IoT) sensor networks, and other emerging technologies can revolutionize monitoring, understanding, and protecting these vital areas. These innovations will enable more precise, real-time monitoring and predictive modeling, allowing for more effective and proactive conservation strategies. The policy recommendations outlined in this chapter provide a comprehensive framework for strengthening KBA protection in India. These recommendations span many areas, including establishing integrated monitoring systems, implementing adaptive management strategies, strengthening legal protections, and integrating KBA conservation into broader land-use planning processes. The emphasis on climate change adaptation, community engagement, international cooperation for transboundary KBAs, and innovative financing mechanisms reflects the multifaceted approach required for effective long-term conservation.

Furthermore, the chapter underscores the importance of capacity building, research promotion, and public awareness campaigns in supporting KBA protection efforts. By investing in these areas, India can build a strong foundation of knowledge, skills, and public support necessary for sustained conservation success. As we look to the future, it is clear that the protection of KBAs will play a crucial role in India's efforts to balance economic development with environmental sustainability. The challenges are significant, ranging from deforestation and habitat fragmentation to the impacts of climate change and increasing urbanization. However, the opportunities for positive change are equally substantial. By implementing the comprehensive policy recommendations outlined in this chapter and leveraging emerging technologies, India can significantly enhance its KBA protection efforts, preserving the country's rich biodiversity and supporting its commitments to global conservation goals and sustainable development.

In brief, protecting Key Biodiversity Areas represents a critical intersection of environmental conservation, sustainable development, and technological innovation. As India continues to navigate the complex challenges of the twenty-first century, effectively managing and conserving these areas will be essential in ensuring a sustainable and biodiverse future for future generations. The path forward requires continued research, adaptive management, cross-sectoral collaboration, and a firm commitment to balancing human needs with preserving India's invaluable natural heritage.

References

1. M.K. Goyal, S. Rakkasagi, S. Shaga et al., Spatiotemporal-based automated inundation mapping of Ramsar wetlands using Google Earth Engine. Sci. Rep. **13**, 17324 (2023). https://doi.org/10.1038/s41598-023-43910-4
2. S. Rakkasagi, M.K. Goyal, S. Jha, Evaluating the future risk of coastal Ramsar wetlands in India to extreme rainfalls using fuzzy logic. J. Hydrol. **632**, 130869 (2024). https://doi.org/10.1016/j.jhydrol.2024.130869
3. R. Kumar, M.K. Goyal, R.Y. Surampalli, T.C. Zhang, River pollution in India: exploring regulatory and remedial paths. Clean Technol. Environ. Policy (2024). https://doi.org/10.1007/s10098-024-02763-9
4. V. Poonia, M. Kumar Goyal, S. Jha, S. Dubey, Terrestrial ecosystem response to flash droughts over India. J. Hydrol. **605**, 127402 (2022). https://doi.org/10.1016/j.jhydrol.2021.127402
5. M. Kumar Goyal, V. Poonia, V. Jain, Three decadal urban drought variability risk assessment for Indian smart cities. J. Hydrol. **625**, 130056 (2023). https://doi.org/10.1016/j.jhydrol.2023.130056
6. S. Bhardwaj, P. Machiwar, C. Kant et al., *Analysis of Urbanization and Assessment of Its Impact on Groundwater and Land Use/Land Cover Using GIS Techniques: A Case Study of Bhopal and Gurugram District* (2023), pp. 219–255
7. S. Singh, M.K. Goyal, E. Saikumar, Assessing climate vulnerability of Ramsar wetlands through CMIP6 projections. Water Resour. Manag. **38**, 1381–1395 (2024). https://doi.org/10.1007/s11269-023-03726-3
8. D. Nigel, J.L. Boucher, A. Cuttelod et al., *Applications of Key Biodiversity Areas: End-User Consultations* (2014)
9. NITI Aayog, *National Consultation on SDGs—Sustaining Life: Integrating Biodiversity Concerns, Ecosystems Values and Climate Resilience in India's Planning Process Focus on SDG 13, 14 and 15* (2017)
10. S.R. Subramoniam, S. Ravindranath, S. Rakkasagi, H. Ram, *Water Resource Management Studies at Micro Level Using Geospatial Technologies* (2022), pp. 49–74

11. V. Gupta, S. Rakkasagi, S. Rajpoot et al., Spatiotemporal analysis of Imja Lake to estimate the downstream flood hazard using the SHIVEK approach. Acta Geophys. (2023). https://doi.org/10.1007/s11600-023-01124-2
12. J. Cavender-Bares, F.D. Schneider, M.J. Santos et al., Integrating remote sensing with ecology and evolution to advance biodiversity conservation. Nat. Ecol. Evol. **6**, 506–519 (2022). https://doi.org/10.1038/s41559-022-01702-5
13. Z. Waliczky, L.D.C. Fishpool, S.H.M. Butchart et al., Important Bird and Biodiversity Areas (IBAs): their impact on conservation policy, advocacy and action. Bird Conserv. Int. **29**, 199–215 (2019). https://doi.org/10.1017/S0959270918000175
14. J. Xue, B. Su, Significant remote sensing vegetation indices: a review of developments and applications. J. Sens. **2017** (2017)
15. B. Majidi, O. Hemmati, F. Baniardalan et al., Geo-spatiotemporal intelligence for smart agricultural and environmental eco-cyber-physical systems, in *Enabling AI Applications in Data Science*. ed. by A.-E. Hassanien, M.H.N. Taha, N.E.M. Khalifa (Springer International Publishing, Cham, 2021), pp.471–491
16. Goyal, C.S.P. Ojha, D.H. Burn, Machine learning algorithms and their application in water resources management, in *Sustainable Water Resources Management*, pp. 165–178
17. S. Singh, D. Kumar, A. Kumar, A. Kuriqi, Entropy-based assessment of climate dynamics with varying elevations for hilly areas of Uttarakhand, India. Sustain. Water Resour. Manag. **9**, 130 (2023). https://doi.org/10.1007/s40899-023-00914-2
18. N. Kumar, M.K. Goyal, Projected changes in monsoonal compound dry-hot extremes in India. Atmos. Res. **310**, 107605 (2024). https://doi.org/10.1016/j.atmosres.2024.107605
19. S. Singh, A. Kumar, Understanding the intricacies of rainfall dynamics using entropy measures. J. Water Clim. Change (2024). https://doi.org/10.2166/wcc.2024.350
20. S. Singh, D. Kumar, D.K. Vishwakarma et al., Seasonal rainfall pattern using coupled neural network-wavelet technique of southern Uttarakhand, India. Theor. Appl. Climatol. **155**, 5185–5201 (2024). https://doi.org/10.1007/s00704-024-04940-8
21. P. Patle, P.K. Singh, S. Rakkasagi et al., *Application of Water Accounting Plus Framework for the Assessment of the Water Consumption Pattern and Food Security* (2023), pp. 257–269
22. K.S. Rautela, S. Singh, M.K. Goyal, Aerosol atmospheric rivers: patterns, impacts, and societal insights. Environ. Sci. Pollut. Res. (2024). https://doi.org/10.1007/s11356-024-34625-8
23. K.S. Rautela, S. Singh, M.K. Goyal, Resilience to air pollution: a novel approach for detecting and predicting aerosol atmospheric rivers within earth system boundaries. Earth Syst. Environ. (2024). https://doi.org/10.1007/s41748-024-00421-0
24. V. Poonia, M.K. Goyal, B.B. Gupta et al., Drought occurrence in different river basins of India and blockchain technology based framework for disaster management. J. Clean. Prod. **312**, 127737 (2021). https://doi.org/10.1016/j.jclepro.2021.127737
25. IUCN, *Guidelines for Using a Global Standard for the Identification of Key Biodiversity Areas: Version 1.2* (IUCN, International Union for Conservation of Nature, Gland, Switzerland, 2022)
26. S. Rakkasagi, M.K. Goyal, Assessing risk levels of the extreme rainfalls in Ramsar wetlands of India using fuzzy logic, in *Chapman Conference on Remote Sensing of the Water Cycle* (AGU, 2024)
27. D.C. Marvin, L.P. Koh, A.J. Lynam et al., Integrating technologies for scalable ecology and conservation. Glob. Ecol. Conserv. **7**, 262–275 (2016). https://doi.org/10.1016/j.gecco.2016.07.002
28. R. Critchlow, A.J. Plumptre, B. Alidria et al., Improving law-enforcement effectiveness and efficiency in protected areas using ranger-collected monitoring data. Conserv. Lett. **10**, 572–580 (2017). https://doi.org/10.1111/conl.12288
29. S.L. Maxwell, V. Cazalis, N. Dudley et al., Area-based conservation in the twenty-first century. Nature **586**, 217–227 (2020). https://doi.org/10.1038/s41586-020-2773-z
30. S. Baidya, P. Chakraborty, S. Rakkasagi et al., Pathways to build resilience toward the impact of climate change on the Indian Sunderban, in *Ecosystem Restoration: Towards Sustainability and Resilient Development*. ed. by A.K. Gupta, M.K. Goyal, S.P. Singh (Springer Nature Singapore, Singapore, 2023), pp.307–333

References

31. V. Jain, K.S. Rautela, M.K. Goyal, *Ecological Restoration: An Overview of Science and Policy Regime* (2023), pp. 1–27
32. J.C. Ingram, D. Wilkie, T. Clements et al., Evidence of payments for ecosystem services as a mechanism for supporting biodiversity conservation and rural livelihoods. Ecosyst. Serv. **7**, 10–21 (2014). https://doi.org/10.1016/j.ecoser.2013.12.003
33. M. Vasilijević, K. Zunckel, M. McKinney et al., *Transboundary Conservation: A Systematic and Integrated Approach* (International Union for Conservation of Nature, 2015)
34. A. Borawska, The role of public awareness campaigns in sustainable development. Econ. Environ. Stud. **17**, 865–877 (2017). https://doi.org/10.25167/ees.2017.44.14
35. Y. Wu, L. Xie, S.-L. Huang et al., Using social media to strengthen public awareness of wildlife conservation. Ocean Coast. Manag. **153**, 76–83 (2018). https://doi.org/10.1016/j.ocecoaman.2017.12.010
36. M.K. Goyal, A.K. Gupta, S. Jha et al., Climate change impact on precipitation extremes over Indian cities: non-stationary analysis. Technol. Forecast. Soc. Change **180**, 121685 (2022). https://doi.org/10.1016/j.techfore.2022.121685
37. M.K. Goyal, A.K. Gupta, J. Das et al., Heatwave magnitude impact over Indian cities: CMIP 6 projections. Theor. Appl. Climatol. **154**, 959–971 (2023). https://doi.org/10.1007/s00704-023-04599-7
38. S. Rakkasagi, V. Poonia, M.K. Goyal, Flash drought as a new climate threat: drought indices, insights from a study in India and implications for future research. J. Water Clim. Change (2023). https://doi.org/10.2166/wcc.2023.347

The manufacturer's authorised representative in the EU is Springer Nature Customer Service Centre GmbH, Europaplatz 3, 69115 Heidelberg, Germany. If you have any concerns regarding our products, please contact ProductSafety@springernature.com

Printed and bound by CPI Group (UK) Ltd, Croydon, CR0 4YY

25/03/2026

02078170-0008